HERBS OF THE BIBLE
And How to Grow Them

HERBS OF THE BIBLE

And How to Grow Them

Allan A. Swenson

CITADEL PRESS
Kensington Publishing Corp.
www.kensingtonbooks.com

CITADEL PRESS BOOKS are published by

Kensington Publishing Corp.
850 Third Avenue
New York, NY 10022

Copyright © 2003 Allan A. Swenson

All rights reserved. No part of this book may be reproduced in any form or by any means without the prior written consent of the publisher, excepting brief quotes used in reviews.

All Kensington titles, imprints, and distributed lines are available at special quantity discounts for bulk purchases for sales promotions, premiums, fund-raising, educational, or institutional use. Special book excerpts or customized printings can also be created to fit specific needs. For details, write or phone the office of the Kensington special sales manager: Kensington Publishing Corp., 850 Third Avenue, New York, NY 10022, attn: Special Sales Department, phone 1-800-221-2647.

CITADEL PRESS and the Citadel logo are Reg. U.S. Pat. & TM Off.

INSERT PHOTOS: Aloe, Chamomile, Fennel, Garlic, Horehound, Majoram, Parsley, Savory, and Thyme photos are courtesty of Cathedral Church of St. John the Divine; Dandelion, Endive, and Melon photos are courtesy of W. Atlee Burpee Co.; Dill and Lavender photos are courtesy of All-America Selections; Onion photo is courtesy of the National Garden Bureau; Saffron Crocus photo is courtesy of Netherlands Flower Bulb Information Center; Sage photo is courtesy of Neot Kedumim. All other photos are from the author's personal collection.

First printing: May 2003

10 9 8 7 6 5 4 3 2 1

Printed in the United States of America

Library of Congress Control Number: 2002113470

ISBN 0-8065-2423-5

Contents

Dedication	vii
Acknowledgments	ix
Prologue: The Wonders of Biblical Herbs	xi

1. Herbs of the Bible: A Scriptural Heritage — 1
2. Biblical Herbs: Tasteful Gifts for Family and Friends — 8
   Aloe, Chamomile, Coriander, Cumin, Dill, Frankincense, Hyssop, Marjoram, Myrrh, Sage, and Wormwood
3. Biblical Herbs: The Bitter Herbs and Food Herbs — 43
   Chicory (Endive), Cucumber Dandelion, Garlic, and the Food Herbs including Leeks, Lettuce, Melons, and Onions—with historic facts, growing, harvest, and use tips
4. Flavor Your Life with Kitchen Herbal Gardens — 70

Anise, Balm, Borage, Caraway, Chives, Comfrey, Fennel, Horseradish, Lavender, Lovage, Marrubium (Horehound), Mint, Mustard, Oregano, Parsley, Pot Marigold, Rosemary, Rue, Saffron, Savory

5. Pick the Right Site, Prepare Well, and Plant Wisely	116
6. Grow Biblical Herbal Tastes Indoors	126
7. Cultivate God's Beauty Everywhere	131
8. Biblical Gardens to Visit	143
9. World's Most Magnificent Biblical Garden	162
10. Biblical Plant Celebration Service	174
11. Favorite Biblical Garden Websites and Plant Sources	184
12. Greatest Global Gardening Sources and Favorite Garden Catalogs	194
Glossary	201
Bibliography	205
Index	209

Dedication

This book is dedicated to my inspiration in life, my beloved wife, Sheila, who with wit, patience, and wisdom has guided me for more than four decades and helped me in every way through more than fifty published books as a guiding light, kindly editor, and lifelong friend. With thanks for many wonderful years, four fine children, and your enduring love.

Acknowledgments

Thanks are due to many marvelous people who have devoted countless hours in answering questions, sharing their knowledge of the Scriptures and gardening, and providing worthwhile and helpful guidance for this book. Special thanks to the Reverend E. Lamar Robinson for his quiet counsel and to Reverend Marsh Hudson-Knapp for his thoughtful suggestions and kind critiques.

Grateful salutes also to all who tend Biblical gardens and graciously shared their knowledge for this book so others will be encouraged to grow Biblical gardens at their houses of worship, in their homes, and in their communities. I was pleased to quote many of you in some of these chapters. My gratitude especially to Page McMahan, Betty Clement, Joseph Scott, Shirley Sidell, Stan Averbach, and other ardent Biblical gardeners for their help.

My heartfelt thanks to the many friends I've met of all faiths and denominations—Protestant, Catholic, Jewish—who offered ideas, tips, advice, and down-to-earth Biblical gardening know-how. May you and your gardens grow gloriously for many years to come.

PHOTO AND ART ACKNOWLEDGMENTS

With special thanks for their help during research:

Peter J. Swenson, J. Drayton Hastie, Nogah Hareuveni, Helen Frenkley, Paul Steinfeld, and Joseph Scott.

—Allan A. Swenson

Prologue

The Wonders of Biblical Herbs

Plants link us to the living reality of the Bible. As you read through the Scriptures you find many references to gardens and flowers, fruits, and other plants. You'll also see mention of herbs. Their first reference is, in fact, in Genesis 1:11: "And God said, Let the earth bring forth grass, the herb yielding seed . . . and God saw that it was good." There are dozens more readings to consider, from the Garden of Eden to Gethsemene.

All of us share this wonderful world that He created. We enjoy the blooming beauty of glorious flowers, the flavors of tasty herbs, the nutrition of fruits and foods, and glory of trees and other plants that grace the earth. We are and must be stewards of His land.

Herbs themselves have been one of God's special blessings. They have provided people with their flavors, fragrances, and medical benefits through the ages. There is so much to learn about herbs, not only those mentioned in the Scriptural passages, but also those that trace their roots to the Holy Land. As you select and grow them you will be taking an adventure into antiquity, experiencing that real link to your ancestors.

Herbs have, of course, been part of human life since written records began. Because they have played such an important role in the lives of people of faith for so many centuries, it is fitting that we focus on them for our own gardens. Whether you plan a Biblical garden that includes herbs, flowers, fruits, trees, and other plants at your place of worship, or just some herbs in your home garden, you'll find joy in discovering them. A garden is essential to a person's life. They're gathering places for all people. Jesus and his disciples often went to gardens to rest, meditate, and pray. They were meeting places where the common folks could gather to share their thoughts and think.

In Deuteronomy 11:10 the children of Israel are given a promise of great blessings that are in store for them and herbs are prominently mentioned. "For the land, whither thou goest in to possess it, is not as the land of Egypt, from whence ye came out, where thou sowedst thy seed and wateredst it with they foot, as a garden of herbs."

In 1 Kings 21:2 we read: "And Ahab spake unto Naboth, saying, Give me thy vineyard, that I may have it for a garden of herbs, because it is near unto my house: and I will give thee for it a better vineyard than it: or, if it seems good to thee, I will give thee the worth of it in money." No doubt Ahab wanted an herb garden conveniently located near his abode with herbs readily available for use in cooking. This could have been the forerunner of today's kitchen herb gardens.

One passage I found myself mentioning on the 700 Club and at Biblical plant talks at the Cathedral Church of St. John the Divine concerned herbs from Numbers 9:11: "The fourteenth day of the second month at even they shall keep it, and eat it with unleavened bread and bitter herbs." That lesson and the instructions in the passages are observed to this day by millions who read the Old Testament. Another admonition is included in Exodus 12:8: "And they shall eat the flesh in that night, roast with fire, and unleavened bread; and with bitter herbs they shall eat it."

Don't let anyone fool you. Herbs as mentioned in the Scriptures are often truly herbs; what many people don't realize is that the term "herb" is more broadly encompassing than we imagine. What we consider today as vegetables or fruits are in fact sometimes considered botanically as herbs. Remember that there are no mentions of vegetables in the Bible but many of herbs.

Reading other passages we note in Psalms 45:8: "All thy garments smell of myrrh, and aloes, and cassia, out of the ivory palaces, whereby they have made thee glad." In the Song of Solomon 4:14, spices that are most often obtained from herbs are again emphasized: "Spikenard and saffron, calamus and cinnamon, with all trees of frankincense; myrrh and aloes, with all the chief spices."

In Biblical times, herbs were especially important for their part in preserving and flavoring food. Today we have refrigeration, freezing, and canning to preserve food. During the times of the Scriptures, people relied on drying, which was a common practice in that part of the world. Salting and smoking were other food-preserving methods.

Because they were so useful and valued, herbs were bartered as part of trade and in taxes. We learn that in St. Matthew 23:23: "Woe unto you, scribes and Pharisees, hypocrites! For ye pay tithe of mint and anise and cummin, and have omitted the weightier matters of the law, judgement, mercy and faith: these ought ye to have done, and not to leave the other undone."

Today many people are focused on healthier diets without fats or salts or unwanted preservatives. Herbs offer us all a marvelous way to flavor and savor food in our own time, a key fact that accounts for the resurgence of herb gardening throughout America. You, too, can join others who are discovering the wonders of herbs.

With this resurgence many more people are discovering the zest to life and eating that herbs can contribute. Even a pinch of one herb or leaf of another can make meals come alive. You can sprinkle herbs on salads or add them to soups. They blend well to flavor meats, fish, and vegetables, too. Herbs today are as versatile and useful as they were in ancient times. Most herbs are easy to grow, are hardy, and don't really require much care. Many also bear beautiful blooms and have marvelous fragrance for decorating and scenting our homes.

And don't forget the chance herbs allow all of us to grow with God, which is something that I've been thinking and writing about for years. It is my hope that you, your children, family, friends, and neighbors will form a deeper relationship with Him by planting, tending, and reaping the rewards of Biblical plants and this book can be part of that.

                                      Allan A. Swenson
                            Windrows Farm, Kennebunk, Maine

Chapter One

Herbs of the Bible: A Scriptural Heritage

> And God said, Let the earth bring forth grass, the herb yielding seed, and the fruit tree yielding fruit after his kind, whose seed is in itself, upon the earth: and it was. And the earth brought forth grass, and herb yielding seed after his kind, and the tree yielding fruit, whose seed was in itself, after his kind: and God saw that it was good. And the evening and the morning were the third day.
>
> —*Genesis 1:11–13.*

Some Scriptural passages, like this story of the creation, emphasize that herbs were placed upon Earth by God for use by man. Other passages relate the importance of herbs to the people of the Holy Land and discuss herbs with regard to feasts and holy observances. Herbs were indeed a vital part of life in Biblical times as they are today.

An overview of Holy Land history helps establish a time line for us. The Early Bronze Age or first historical period discussed in the Bible extended from 3300–200 B.C. when histories tell us the first states emerged. Egypt had its early dynastic period from 3000–2700 B.C. Then the so-called Pyramid Age extended from

2700–2000 B.C. During that time there were close commercial relations by sea and land with Syria and Palestine.

In nearby Babylonia the first dynastic age, called the Sumerian Age, extended from 2800–2400 B.C. During this period the world's first great empire was founded. Following this, the so-called Middle Kingdom in Egypt extended from 1990–1778 B.C. and Palestine and Syria became part of it. Then Asiatic rulers, probably Canaanites, seized control of Egypt and built an empire that included the Egyptian possessions.

The New Kingdom began when Egyptian rulers took control. Names such as Thutmose, 1490–1435 B.C., Amenhotep III, and Rameses II (1290–1224 B.C. who fought the battle with the Hittites in Syria circa 1270 B.C.) are associated with this period. Then Merenptah is credited with defeating Israel in Palestine circa 1220 B.C., which leads to Moses and the Exodus of Israel occurring at approximately 1290 B.C.

Meanwhile in that famed "fertile crescent" of Babylonia, war was a periodic plague. It has been recorded that a Hittite army raided Babylon circa 1550 B.C. and conquered northern Syria. Ultimately, about 1270 B.C., a peace treaty was established with the Egyptians. As these armies moved about they brought with them foods and herbs. During the period 1500–1200 B.C., Palestine was nominally a province of Egypt, which is documented in letters from Canaanite kings to the Egyptian court.

Elsewhere during this period, the Golden Age of Crete reached its peak and Greek influence spread throughout the Mediterranean. Palestine also was seeing its earliest development. The beginning of the Iron Age, 1200–1000 B.C., saw important changes in Palestine. It was the period of the Judges, circa 1200–1020. The Philistines were pushed back from Egypt and settled along the southern coast. Also the great battle with Canaanites occurred in the Valley of Jezebel as noted in the Song of Deborah. This also was the time that included the struggle with the Philistines, the fall of Shiloh, the death of Eli circa 1050 B.C., and the beginning of Israelite monarchy under Saul, 1020–1000.

Following were more war and strife. Ezekiel was exiled and Nehemiah returned circa 445 B.C. The Roman Empire was expanded and in 63 B.C. Pompey established Roman control over Palestine. During the Roman period, 63 B.C. to approximately A.D. 325, wider trade flourished. Herbs from other areas were undoubtedly introduced.

Palestine has had a remarkable influence and place in world history through the centuries. It's actually a small area considering its impact on the world. Not only is the area a Holy Land to three religions, but also a crossroads of caravans and trade for the millennia. Today the area blooms and blossoms again thanks to the dedication of farmers who understand the soil, climate, microclimates, and plant needs for peak productivity.

West of the River Jordan the total area of the Holy Land is only 6,000 square miles. Adding the portion east of the Jordan there are only 10,000 square miles in the entire region. Put into context here, the whole region is barely larger than the small state of Vermont.

The Israelites spoke of their country as extending from Dan to Beersheba, which we read about in 1 Kings 4:25. From the Dan river at the foot of Mt. Hermon to Beersheba where the hill country ends and desert begins is somewhat less than 150 miles. East-west distances are even smaller. From Accho on the coast to the Sea of Galilee is only 28 miles. Yet, despite the size, the Holy Land is remarkable for its climate and plant-growing diversity. Understanding these differences will help gardeners who are challenged by different growing conditions in their own areas. Fact is, there are perhaps more different climates and microclimates in that part of the world than almost anywhere else.

Let's look at the "climate and plant-growing conditions" in these lands. Along the coastal plain a mild climate exists and only 35 miles away inland at about 2,600 feet elevation is Jerusalem with a temperate climate. East of the Holy City, only 15 miles away at Jericho, is actually 3,400 feet below Jerusalem and nearly 800 feet lower than the actual level of the Mediterranean Sea. In that area a virtual tropical climate is present. Summer heat can be intense. A bit farther east the Transjordan plateau reaches an altitude of 4,000 feet and snow falls in winter.

From Mt. Herman at 9,000-plus feet with it alpine range and

snowcover, the land slopes into the Jordan valley at sea level. The Sea of Galilee, which is nearly 700 feet below sea level to the Dead Sea, 1,290 feet below sea level, has mild climates throughout the year. These different climatic conditions—indeed, microclimates—depend on area, altitude, land forms, and other factors, including prevailing winds. Think about this with regard to your own property. These key points also apply to your own landscaping. During the King Solomon era in the Holy Land there was a special emphasis on nature and plants, as we know from his famed garden. More plants and plant products, actually 34 of them, are associated with Solomon than with any other Biblicized person. They range from fruits and foods to flowers and herbs. Biblical scholars credit two people mentioned in the Bible as being botanists. One is Jotham who delivered a lecture about the plants of Mount Gerizim in Judges 9:1. The other is Solomon who was the son of the warrior King David. Solomon reigned from approximately 1015–975 B.C., which was a glorious period of prosperity as described in 2 Chronicles 9:13–28. Solomon's accomplishments included his writing the "Songs of Solomon" as well as his encouragement of culture and the construction of a temple and palaces.

In the Song of Solomon, a rather short book of only eight chapters, you can find the plant references. Among them is hyssop, which is also referred to elsewhere in both the Old and the New Testaments. It is worth reading John 19:29–31, and Hebrews 9:1 about it as well.

Hyssop, actually a product of the hyssop plant, was an important part of the Passover, as indicated in Exodus 12:22. It was used as a ceremonial cleansing from skin disease, which is mentioned in Leviticus 14. David also mentions hyssop in Psalm 51:7; a New Testament reference is in John 19:29. What is the hyssop the Bible is actually referring to? Probably *Origanum syriacum L.*, a plant known in English as Syrian hyssop and a relative of the well-known kitchen herbs oregano and marjoram. Even today, *Origanum syriacum* is one of the most widely used and valued herbs of the Palestinians.

Of course we have to mention myrrh as an important herb of

the Bible. One of the best-known, it is the dried resin of several species of Commiphora, which are shrubs or small trees. They grow in the arid and semiarid regions of East Africa, Arabia, and the Indian subcontinent. Botanical scholars say that "myrrh" may have had opiate qualities. If so, that enables us to better understand Mark 15:23 where Jesus, on the cross, was offered vinegar mingled with myrrh but refused it. As Biblical scholars explain, He would not be drugged.

There are two different myrrhs, medicinal and fragrant, and they are both translated from the same Hebrew word, *mor*. References to the scent of myrrh can be found in the pages of Solomon's writings. For example, I found six references to myrrh in the Song of Solomon; no other Bible author mentions it as much.

Also mentioned in the Song of Solomon is saffron, a spice that comes from the saffron crocus flower, *Crocus sativus L.* (Thousands of saffron flowers are needed to produce 1 gram of this spice. Next time you are in the grocery store, look for saffron. Now you can understand why it is so expensive.) The purple crocus flowers are as lovely today as they must have been in the past, peeping through and welcoming spring in our gardens. A most striking feature of the flower is its large, drooping stigmata. "My beloved is unto me a cluster of henna flowers in the vineyards of En Gedi" (Song of Solomon 4:14).

Returning briefly to Biblical plant history, visualize what you once learned in school about Babylon. It was the apex of civilization from the earliest time to the golden age of Greece. Mesopotamia (Babylon) and Egypt were the cultural forces then. As gardeners can understand, water was the key. These earliest civilization centers grew and prospered because they tapped the Tigris, Euphrates, and Nile rivers to provide early farmers and gardeners with water for their crops.

As you read the Scriptures you'll find other allusions to herbs and their prominent place in the history, heritage, and life of the people during Biblical times. One example is Abraham and Jacob with regard to their family's migration in Genesis 11:31.

In Deuteronomy, the children of Israel are given a promise of great blessings in store for them. Herbs are prominently mentioned.

"For the land, whither thou goest in to possess it, is not as the land of Egypt, from whence ye came out, where thou sowest thy seed, and waterest it with thy foot, as a garden of herbs" (Deuteronomy 11:20). Later in the Scriptures herbs are emphasized again as Ahab spoke to Naboth. According to 1 Kings 21:2: "And Ahab spake unto Naboth, saying, Give me thy vineyard, that I may have it for a garden of herbs, because it is near unto my house: and I will give thee for it a better vineyard than it; or, if it seem good to thee, I will give thee the worth of it in money."

Today, we think of herbs with regard to exotic and pleasant aromas. Kitchen herb gardens are making a deserved comeback. People are growing a wide variety of herbs in herb gardens, mixed into vegetable patches, using flowering herbs in flower beds and borders and cultivating herbs in pots, tubs, and buckets as container gardens. From these herb sources they can snip, pick, and pluck the herbs they want for more delicious meals. Good for them. Good for you, too, as you learn in this book about the many values of herbs and ways to grow them and use them for tastier living. Ahab seems to have tended the same type garden of herbs conveniently located near his house and readily available for cooking in those Biblical days.

Among the herbs mentioned in the Scriptures are ones we know with a fair amount of accuracy. By tracing their roots, observing growing similarities with plants still thriving in the Holy Land, and reviewing translations of Scriptures, Biblical and botanical scholars have come to an agreement on the contemporary names given the herbs discussed in the Bible. For example, aloe and coriander, frankincense and myrrh, sage and wormwood are easily identified. Others have required serious interpretation to confirm Biblical references. Even the "bitter herbs" can be identified with a fair amount of certainty. Most authorities agree that these included lettuce, chicory, endive, leeks, and onions.

It is appropriate to focus on some of the thoughts by the

Biblical and botanical scholars, from monasteries and convents to those who have meticulously researched Biblical herbs and focused on Scriptural references.

While writing, I have integrated their thoughts and heeded their growing advice (some have tended Biblical gardens around the United States and the world). For convenience, I've grouped the herbs so you can find those with Scriptural references, those that are most likely the bitter herbs mentioned in Exodus 12:8 and Numbers 9:11, and those that are found naturally and grown extensively in the Holy Land. Some of these may have a Scriptural mention while others are simply known to grow in the area—and have grown there since the time of the New Testament if not before—so I believe they deserve a place in herb gardens and meal enhancement.

Because some of the herbs are especially attractive for their floral displays, they can absolutely be used as part of your home landscape—not just as "herbs" in an herb garden. Fact is, many more gardeners are adopting flowering herbs for their beauty. Oftentimes, herb gardeners snip and pick and pluck herbs for use in cooking, which tends to leave an herb garden looking somewhat ragged. That has led to the trend to use more herbs in plantscapes, a good idea whose time has come. In outdoor entertainment areas they also add fragrant aromas no home garden should be without.

Chapter Two

Biblical Herbs: Tasteful Gifts for Family and Friends

This chapter discusses the best-known herbs that have Scriptural quotations and passages. In addition, I've included the best how-to-grow tips, ideas and advice from my own years of gardening combined with the input from many devoted Biblical herb growers.

Herbs are usually divided into several groups by old-time herbalists, some of them overlapping. The romantic herbs are those grown for fragrance, the culinary herbs for flavoring in cooking, the medicinal herbs for use in medicines and health, and there are also herbs grown for making colors and dyes. Let's look at them.

ALOE: *ALOE VERA*

> And there came also Nicodemus, which at the first came to Jesus by night, and brought a mixture of myrrh and aloes, about an hundred pound weight. Then took they the body of Jesus, and wound it in linen clothes with the spices, as the manner of the Jews is to bury.
> —St. John 19:39–40

The aloes of the New Testament, according to scholars, *(Aloe vera)* were succulent plants with long sword-like leaves with serrations and erect flower heads up to 3 feet high. Biblical scholars have noted that, historically, the bitter pith was used as a medicine and also for embalming.

There has been, according to scholars, a difference between the aloes of the Old Testament and the New Testament. The aloes of the Old Testament, after studying the context in which they are mentioned, could possibly have been trees. Reviewing Greek translations as well as old Hebrew texts, some authorities have claimed the eaglewood tree, *Auilaria agalocha,* was the Old Testament aloe. In fact, the context in the Douay, Moffatt, and Goodspeed translations of the Bible seem to indicate a tree. The Moffatt translation actually substitutes the word "oaks" for aloes.

However, it must be remembered that many scholars throughout the centuries have translated the Bible without knowledge of botany and plant physiology. Although today some believe the aloe to be eaglewood or even sandalwood trees, *Santalum album,* neither of these trees are native to the Holy Land.

Because of the debate, you may wish to refer to some of the Scriptural passages in Psalms 45:8. You will read: "All thy garments smell of myrrh, and aloes, and cassia," and in Proverbs 7:17: "I have perfumed my bed with myrrh, aloes, and cinnamon." In both cases, aloe appears grouped with other perfumes, scents, and spices. The aloes indeed seem to be a type of plant that yields resinous gum from which incense is made. It's a difficult call.

With those considerations in mind, the aloes of the New Testament mentioned by St. John deserve closer attention. It is known from historical sources that the ancient Egyptians had perfected the art of embalming to a high state. The juice of the true aloe, *Aloe succotrina,* was known to the Egyptians and it is likely that it was the material used in their embalming art. That seems to provide a better understanding of the mention of aloe brought by Nicodemus to anoint the body of Jesus. Maybe we'll never be able to decode the answer to this Biblical allusion.

Aloe history dates back far. Greek writings note the use of the plant as a healing herb more than 2,000 years ago. Supposedly

Alexander the Great valued aloe as the herb needed to heal wounds for his troops. Research shows that people throughout the world have been aware of aloe's properties, and in some areas the gel of aloe is used to heal burns and help people look younger.

The *Aloe succotrina* has a thick stalk with fleshy leaves—as do all succulents. Its smell is odd and its taste bitter. Although the plant is a native of the desert areas of east coast Africa, it is strikingly similar to other aloes that can be grown as houseplants. Since it is not practical and nearly impossible to obtain the exact species to duplicate every plant of the Scriptures, we may need to be satisfied with our closest substitutions.

In native habitats some aloes grow almost tree-like (and this may be what the Old Testament was referring to). Few homes could accommodate such huge specimens. Close relatives make good family representatives. Most aloes are sun lovers, so the best place to grow them is outdoors on the patio or balcony in summertime. When placing aloes outdoors, acclimate them to full sun or they may burn. Sunny windows are best for indoor cultivation.

A wide choice of aloes gradually exists for the home grower: from the spiny aloe, *A. africana,* to the tree aloe, *A. Arborescens.* This tree-like plant has branching stems that carry spreading rosettes of tapering leaves. In winter sun it often blooms with vermilion to yellow flowers in spiky clusters. It is easy to grow in sandy soil mix but is better suited to arboretums. It can be propagated by making top cuttings or rooting suckers.

Aloe vera chinensis is an Indian medicinal aloe, a small Asiatic form with fleshy, lanceloate leaves that curve at the tips and are bluish green with white markings. That and the *Aloe vera,* or medicine plant aloe, are easily grown on windowsills. *Aloe vera* is a short stemmed freely suckering plant with sword-shaped leaves that may be bluish or grayish green and grow in a rosette pattern. The juicy pulp of this variety makes the balm used for a poultice to heal burns and cuts in its native areas of Africa.

Depending on how exact you wish to be in closely duplicating herbs that are closely native to Holy Land plants, many

Biblical gardeners allow themselves some latitude. Dozens of aloe cultivars have been developed by nurseries and plant breeders. Best propagation is achieved by dividing or planting cuttings or suckers that form on the aloe plants.

It is helpful to remember that the true "aloe" should not be confused with another plant, the agave or American aloe, that is, *Agave Americana*. Although often popularly called "aloes," these are plants of the New World and not native to the Holy Land.

Plant Profile

Aloes typically have a rosette of long, tapering, and fleshy-type leaves. When broken they exude a sticky sap. There are about 300 species of these perennial plants, and they range from being short to some with 15-foot stalks.

Growing Tips Use sandy loam for aloes. Place in sunny locations and water sparingly, at least once a month. They can tolerate some shade. Propagate from suckers that sprout around the plant base. Once established, aloe plants can grow well for many years and even seem to like crowded conditions. Authorities suggest using older, outer leaves for their gel.

CHAMOMILE: *CHAMAEMELUM NOBILE* AND *MATRICARIA RECUTITA*

> A voice says, "Cry." And he said, "What shall I cry? All flesh is grass and all the goodliness thereof is as the flower of the field. The grass withereth, the flower fadeth: because the spirit of the Lord bloweth upon it: surely the people is grass. The grass withereth, the flower fadeth: but the word of our God shall stand forever."
>
> —Isaiah 40:6–8

Not only does chamomile have its roots in the Holy Land, but this herb traces to Roman times. Indeed, *Anthemis nobilis* is known as Roman chamomile, a low-growing perennial. If you remember your children's stories, you'll probably recall that Peter Rabbit's mother brewed him a cup of chamomile tea. Even today popular teas promote its ability to relax the overstressed.

Tracing this herb back to antiquity, historians report that chamomile was reportedly used in early Egyptian times to cure malarial chills that were common in that part of the world. The famed herbalists Discorides and Pliny wrote about this herb as a remedy to relieve headaches. Researching further we find that chamomile tea has been enjoyed for centuries around the world, from medieval England to Spain, where it has also been used to flavor wine. In Spain it is called "manzanilla," or "little apple" in rough translation, and there the chamomile flowers have long been used as a flavoring for light sherry wine.

Tracing this herb's history further, it has been used as a tea for the purpose of lightening hair or accentuating blond highlights according to old recipes. Today, it is used in products from both the perfume and the cosmetics industries. Some also report that chamomile has been used to help eliminate odors from unrefrigerated meats and even as an insect repellent.

For years this attractive herb has been used in potpourris where the apple fragrance adds a delightful aroma. The plant has also been popular for dried flower arrangements. In France, chamomile is often used as an after-dinner drink mixed with mint in place of pekoe tea or coffee. While traveling in England and Ireland I discovered this low-growing herb growing along walkways. When the plants are crushed they give off their distinctive aroma and don't seem to mind being walked on—or even mowed.

Although chamomile has one name, few gardeners or others realize that in reality this is two different plants: one a perennial and one an annual. Both have distinctive daisy-like blooms, feathery foliage, and the distinctive apple aroma. Obviously there is no relationship to real apples but the fragrance certainly carries a similar scent. Not surprisingly then, chamomile in Greek means "ground apple."

Plant Profile

So-called Roman chamomile, *Chamaemelum nobile,* is a low-growing perennial that normally grows 7 to 9 inches tall. This type of chamomile has a solid central disk of deep yellow color and rays or petals of white to cream growing at the end of downy stems. The plant was once known by another Latin name, *Anthemis nobilis.* Some mail-order catalogs may still continue with that former name so be on the lookout when you shop by mail.

So-called German chamomile, *Matricaria recutita,* is an annual plant that grows 2 to 3 feet tall. It, too, has a solitary central disk, yellow color, and petals or rays of silvery white to cream. Blooms appear at the end of downy stems. Leaves are alternate and divided into segments with a feathery appearance. This herb is less fragrant than Roman chamomile. Both flower from late spring through late summer and provide an abundance of appealing blooms. They also are native to the Middle East—plus Europe, Africa, and Asia—and have naturalized in North America where they are widely cultivated.

Growing Tips Roman chamomile is your best bet for gardens primarily because it is a perennial. Once established it returns to reward you with its blooms and benefits every year. It will grow in almost any kind of soil and can do well in sandy loam, but newer cultivars prefer richer soils with more organic matter. Seeds are tiny so just sprinkle on top when planting and cover lightly with a soil mix. Be patient because chamomile is slow to sprout. Most veteran gardeners recommend buying prestarted plants. Roman chamomile is easily propagated from offshoots that the parent plants produce; just reset them in early spring. Next, simply dig up young plants, reset them in their new location, and press soil well around them. Water well until they have set their roots properly. These herbs are reasonably hardy in most parts of America. They also are especially useful as ground cover on slopes and in rock gardens. Grow them in a large pot or tub for year-round use. Put the container outdoors in summer and bring indoors for winter.

German chamomile can be seeded in the spring in a well-prepared bed or area. Cover seeds lightly. Once the plants grow they will often reseed themselves when some flowers form seeds and drop them. Try this type in a container, too. Both types of chamomile prefer lots of sun. Keep soil evenly moist for best growth and flowering.

For potpourris and dried flower arrangements pick the flowers at their peak and hang to dry or spread on clean sheets. To make tea, carefully harvest the flowers for drying when the petals turn back toward the disk. After drying, look up recipes for making tea and blending with other herbs as well. Dyer's chamomile is a slightly taller plant, 14 to 18 inches tall with bold yellow color; it blooms longer than others.

CORIANDER: *CORIANDRUM SATIVUM*

> And the manna was as coriander seed, and the color thereof of bdellium. And the people went about, and gathered it, and ground it in mills or beat it in a mortar, and baked it in pans, and made cakes of it: and the taste of it was as the taste of fresh oil. And when the dew fell upon the camp in the night, the manna fell upon it.
> —*Numbers 11:7–9*

> So the people rested on the seventh day. And the house of Israel called the name thereof Manna: and it was like coriander seed, white; and the taste of it was like wafers made with honey. And Moses said, This is the thing which the Lord commandeth. Fill an omer of it to be kept for your generations; that they may see the bread wherewith I have fed you in the wilderness, when I brought you forth from the land of Egypt.
> —*Exodus 16:31–32*

Coriander is another Scriptural herb that traces its roots and uses far back into the ancient world. Herbalists report that during the Chinese Han Dynasty, circa 207 B.C. to A.D. 220, there were several products made from coriander. Writings indicate that coriander was thought to have the power to make people immortal. Written records about historic events, agricultural crops, and medicinal uses of herbs reveal that coriander probably has been cultivated and used for more than 3,000 years.

Egyptologists have said that coriander seeds have been found in Egyptian tombs where funeral offerings of herbs and foods and other items were placed with deceased royal families. It is interesting to note how widely this herb was used and mentioned in historic writings. Hippocrates supposedly made medicines from it. Others report that coriander was used by both Greek and Roman doctors.

Linguists say that the Hebrew word *gad* is rendered as coriander in most translations of the Scriptures. Most likely this refers to the common coriander plant, which is an annual white or reddish flowered herb about 15 to 20 inches tall with slender stems, deeply cut leaves, and a strong odor. A few Scriptural authorities believe that coriander was perhaps one of the "bitter herbs" involved in the ritual of Passover but most believe lettuce, endive, dandelion, and other leafy herbs were more likely candidates. Nevertheless, coriander is expressly identified in the two Scriptural references that head this herb's section, so we can positively identify it as one of the truly Biblical herbs. If you wish to pursue Scriptural research even further, you can explore mentions of coriander in the Talmud.

From its mention in the Bible and tracing its rightful place through the ages, we realize that coriander has been used as a food for flavoring in soups and salads. It also has been a favored ingredient in hot curries and sauces. To the ancients it was also used as medicine as an aromatic and carminative for indigestion. Coriander commonly grows wild in grain fields and along roads in the Middle East countries, especially in the Holy Land and Egypt. According to Harold N. Moldenke, Ph.D. and Alma L. Moldenke, B.A., authors of the classic *Plants of the Bible,* in olden times a favorite drink was made by steeping coriander plants in wine. Then they were dried and thus rendered milder when eaten with various other foods. Coriander is still enjoyed today in the cuisines of Egypt, Iran, and India.

Coriander fruit, improperly called "seed," comes from the *Coriandrum sativum* plant. The plant produces a slender, erect, hollow stem that rises 1 to 2 feet tall. The fruit itself is globular and smooth with five indistinct ridges. Coriander differs from dill and caraway in its globular pearl-like seeds, which are quite aromatic. The seeds are today used in confectionery. The leaves also are aromatic and are used in soups and to flavor puddings and wines. In fact, coriander is presently used as a spice in many Arabic countries. According to legend, coriander was once thought to arouse passion, and was even mentioned as an aphrodisiac in an Arabian fantasy. The herb is native to the eastern Mediterranean area and southern Europe. It also grows in India, Mexico, other parts of South America, and North America.

Plant Profile

The coriander plant is a bright green annual herb that has slender, erect, finely grooved stems with a shiny, hairless appearance. Flowers are tiny, white to pink to reddish with 5 to 10 rays on short-stalked umbels. Outer flowers tend to be larger than inner ones. Leaves are compound and pinnate. Lower leaves are roundish and lobed. Upper ones are more finely divided into narrow segments. Coriander fruit are brownish yellow, spherical, and ribbed, about ¼ inch long in symmetrical clusters. The plant grows 2 to 3 feet tall.

Flowering occurs in late summer, depending on seeding time. Although the plant has an unpleasant smell when maturing, as fruits ripen the fragrance becomes more like lemony citrus. Because this herb attracts bees and other pollinators it proves of bonus value for gardeners.

Growing Tips Coriander roots are long, so it is difficult to transplant these herbs. Sow seeds ½ inch deep directly in beds when frost danger has passed. Space 6 to 10 inches apart. Coriander prefers a moderately rich, light, and well-drained soil in sun but can tolerate partial shade. Some herbalists suggest companion planting with the biennial caraway, because of similar growth habits. Water regularly and mulch to thwart weeds and retain soil moisture. Avoid over-fertilizing. Experts note that too much nitrogen reduces the flavor of this plant.

Coriander is widely known by the name "cilantro" and is available fresh in many grocery stores. It is popular in making salsa. The Bible doesn't talk about the actual uses of coriander but the appearance of its fruit as a useful comparison to help people picture manna in their minds. Coriander leaves have a strong sage flavor and a bit of a citrus taste, too. It is best to harvest coriander when leaves and flowers have become brown and before the seed has a chance to fall. Cut the plant and hang it to dry over clean sheets. Be aware that undried seeds have a bitter taste. To harvest fresh leaves cut the small immature ones for best flavor. Dried leaves don't store well but seeds will keep nicely in glass jars. Herbalists note that whole or ground seeds give sparkle to marinades and salad dressings, as well as in chili. Enjoy leaves and seeds in soups and stews as well as in salads. Try small bits to test the amount that pleases you.

CUMIN: *CUMMUM CYMINUM*

> Doth the plowman plow all day to sow? Doth he open and break the clods of his ground? When he hath made plain the face thereof, doth he not cast abroad the fitches, and scatter the cummin, and cast in the principal wheat and the appointed barley and the rice in their place? For his God doth instruct him to discretion, and doth teach him. For the fitches are not threshed with a threshing instrument, neither is a cart wheel turned about upon the cummin; but the fitches are beaten out with a staff, and the cummin with a rod.
> —*Isaiah 28:24–27*

> Woe unto you, scribes and Pharisees, hypocrites! For ye pay tithe of mint and anise and cummin, and have omit-

ted the weightier matters of the law, judgement, mercy and faith: these ought ye to have done, and not to leave the other undone.

—Matthew 23:23

Linguists say that the Hebrew word for cumin is *kammon* and no one doubts the identity of this plant. It is said by some botanists to be native to Egypt, although some say its origin is Syria. Still others, perhaps because of cumin's popularity in India, believe it was a native there before being transplanted by caravans throughout the Middle East. Historic documents reveal that in the thirteenth and fourteenth centuries cumin was much in use as a culinary spice in England. Today in Europe, cumin has been replaced by caraway seed, which many concur has a more agreeable flavor. Cumin's principal use today is as an ingredient in curry powder.

The herb actually belongs to the parsley family. It has finely cut leaves and clusters of small white to rose flowers. The fruits—seeds—are used as an ingredient in curry powders, as mentioned, and for flavoring pickles and soups. Today cumin is grown mostly in India, Iran, Indonesia, China, and in the southern Mediterranean.

Harold N. Moldenke writes that in ancient times, Ethiopian cumin was considered to be the best available, followed by that from Egypt. It was used as a condiment with fish and meats, in stews, as an appetite stimulant, and in medicine. Egyptian cooks sprinkled the seed on breads and cakes, Moldenke says, much as we do caraway seeds today. It is interesting to note that the Basic English version of the Bible substitutes "all sorts of sweet-smelling plants" for "mint, and anise and cummin" in the Matthew chapter and verse. All other English versions use 'cumin' or 'cummin' in both Isaiah and Matthew. Black cumin or *Nigella sativum*, which is mentioned in Jastrow's translation of the Bible for cumin, is very likely a correct rendering of the Biblical plant, according to botanist Michael Zohary. He asserted that it is the plant denoted by the Hebrew *ketzah* of Isaiah 28:27. He seems to have based his conclusion on the historic custom of flavoring with it by sprinkling the seed over baked goods as well as the linguistic similarity of the names for nigella in Arabic and Aramaic.

Please note: Not all Bible versions are translations! Paraphrases are versions that put the gist of the Bible into the author's own words. Sometimes this makes good reading but it is questionable for study because only one person concluded that this is what a text means. As we expand our Biblical knowledge it is helpful to use Strong's concordance, which groups words by the Hebrew or Greek word in the Bible. Because some translators use the same English word for two or more Hebrew words, meaning can sometimes be ambiguous, but the Concordance is a big help.

Maude Grieve in her excellent book, *A Modern Herbal*, provides some other fascinating facts about this history of cumin. She notes that "From Pliny we learn that the ancients took the ground seed medicinally with bread, water or wine, and that it was accounted the best of condiments."

Cumin has been a popular spice all over the world, especially in Latin America, North Africa, and Asia. Tracing historic documents we discover that cumin was a common spice in the times of the Roman Empire. Indian cuisine, of course, has made much of cumin where the fruits are fried or roasted. Unfortunately it has lost favor in North America and seed may be difficult to find. Be persistent in your search and you can add this authentic Biblical herb to your garden.

Plant Profile

The cumin plant has a slender stem that is branched and grows to about 1 foot tall. Leaves are divided into long, narrow segments much like fennel but much smaller. They are of a deep green color and generally turned back at the ends. Flowers are small, white to rose colored in stalked umbels with four to six rays that are about ⅓ inch long. Cumin typically blooms in June and July followed by fruit, the so-called seeds in popular terminology. These are oblong in shape and thicker in the middle resembling caraway seeds but lighter in color and bristly instead of smooth. Herbalists note that the smell and taste are somewhat like caraway.

Growing Tips It is best to start cumin from seeds. They sprout and grow rather well and plants respond well when grown in fertile, loamy soil with lots of sunlight. Since seeds are somewhat difficult to obtain your best bet may be herb enthusiasts. Research some herb websites or ask around your area. Start seeds in small pots filled with seed starter mix. Thin to two seedlings per pot and harden off before planting outdoors in spring. Be sure to protect the balls of earth which adhere to the roots in the pots. Use mulch to keep clear of weeds and your plants should flower and produce their seeds well. Harvest seeds by cutting and drying plants, and then thresh out seeds when they are dried.

DILL: *ANETHUM GRAVEOLENS L.*

> Dill is not threshed with a threshing sledge, nor is a cart wheel rolled over cumin; but dill is beaten out with a stick, and cumin with a rod. Grain is crushed for bread, but one does not thresh it forever . . .
> —*Isaiah 28:27* (Goodspeed version)

> For the fitches are not threshed with a threshing instrument, neither is a cart wheel turned about upon the cumin: but the fitches are beaten out with a staff, and the cumin with a rod.
> —*Isaiah 28:27* (King James translation)

It is important to read other translations of the Bible because names of plants, flowers, and herbs can vary depending. To find different versions of the Scriptures log on to www.biblestudytools.net, where you can enter a Scriptural passage and find the same passage in different versions. This can prove useful when trying to track down the herbs of the Bible.

Widely cultivated for its seeds, which are aromatic and similar to those of caraway, dill is used in cooking for flavoring and especially for making pickles. Although most likely a native of Europe, dill grows wild and is cultivated in Palestine and Egypt where it has been known for centuries. In fact, four translators of the Bible—Moffatt, Weymouth, Goodspeed, and Lamsa—all concur on using "dill" in the passage cited above.

Widely used in Scandinavia and Germany for pickling, the actual word "dill" may be related to the Old Norse word *dilla*, which means to soothe or calm. Some linguists believe that may relate to the fact that in olden days dill was used to relieve stomach pain in babies, calming them. Over the years dill has continued its time-honored tradition as a key ingredient in making the famous dill pickles, as well as preserving types of pickled fish in Norse countries. Dill is a hardy annual, an original native of the Mediterranean region and southern Russia according to botanists. It grows wild in Spain, Portugal, and the coast of Italy.

According to Grieve, dill was well known in Pliny's days and is often mentioned by writers in the Middle Ages. She reports that it occurs in the tenth-century vocabulary of Alfric, Archbishop of Canterbury. Dill seems to also have been one of the plants used by magicians and the superstitious during the Middle Ages to cast spells against witches and witchcraft.

A tall plant with finely divided leaves and flowers, dill is an annual, which sets its seeds in late summer and early fall. The herb's feathery, light green leaves are almost fern-like and lend an attractive look to landscapes. Both leaves and seeds are useful, but be prepared to capture the seeds when they ripen before they fall.

Plant Profile

Dill thrives in moderately rich, well-drained, and moist soil in full sun. Typically, dill plants grow 2 to 3 feet high and look much like fennel, although dill is smaller. Unlike fennel, dill seldom produces more than one stalk from its long, spindle-shaped root; the plants have very upright growth. The herb has smooth stems, which are shiny and hollow. By midsummer dill plants produce flat terminal umbels with numerous yellow flowers and the petals are rolled inward. Flowers may be 6 inches across. Leaves are feathery, bluish green with thread-like, pointed leaflets. The flat fruits, or seeds, are ribbed, flattened, and about ⅙ inch long and are produced in great quantities. These seeds are pungent and bitter in taste and very light. There can be many thousands of seeds in one umbel. The entire plant is aromatic.

Growing Tips Fortunately, dill is easy to grow. If you enjoy making pickles, go to it. Sow seeds directly in the ground when frost danger is over. Place 3 to 5 seeds per inch, about ¼ inch deep in sunny locations. Use dill to advantage as it grows for attractive landscape accents, as a backdrop for shorter herbs or flowers, and to attract bees. Remember that dill can be self-seeding, so new plants may emerge from dropped seeds. When planting several rows, space them 2 to 3 feet apart. Because they are tall, avoid windy areas that might damage them or use stakes for support. If you want to use leaves regularly, consider succession plantings every few weeks to ensure fresh, young leaves.

When sprouts are a few inches high, thin seedlings to 5 to 6 inches apart. Be patient with dill, it can seem very slow-growing. Once it begins, however, in outdoor gardens with fertile soil dill grow 3 to 4 feet tall. Indoor plants require a deep container to accommodate the long tap root. Best bet is to grow your pickle-making dill crop outdoors and enjoy its feathery foliage in your landscape.

To enjoy fresh dill, clip tips of the feathery foliage and chop up for use in soups or over boiled meats and fish. Cut leaves are delicious in sandwiches of cream or cottage cheese. To harvest for other uses, cut the best leaves and place them in a warm, dry area. Leave them for several days. After dill is thoroughly dry, crumble it and place it in airtight containers.

Seeds are probably more useful, especially for making pickles. To harvest seeds, be alert! Dill flowers are a greenish yellow and can produce large quantities of seeds, which turn light brown when ripe. To harvest without losing seeds, spread a clean white sheet beneath the plants and shake them to dislodge the seeds. Then cut off seed heads and finish cleaning out seeds. Rub them to remove the chaff and then store seeds in their own glass container after they are fully dried. Or cut plants with leaves and nearly mature seed heads and hang them in a cool, dry place to air dry. But be sure to place a clean sheet beneath them to catch the seeds as they fall.

FRANKINCENSE: *BOSWELLIA SACRA*

And thou shalt speak unto the children of Israel, saying. This shall be an holy anointing oil unto me throughout your generations. Upon man's flesh shall it not be poured, neither shall ye make any other like it after the composition of it: it is holy, and it shall be holy unto you. Whosoever compoundeth any like it, or whosoever putteth any of it upon a stranger, shall even be cut off from his people. And the Lord said unto Moses, Take unto thee sweet spices, stracte, and onycha, and galbanum; these sweet spices with pure frankincense: of each shall there be a like weight.
—*Exodus 30:31–34*

And thou shalt put pure frankincense upon each row, that it may be the bread for a memorial, even an offering made by fire unto the Lord.
—*Leviticus 24:7*

When they saw the star, they rejoiced with exceeding great joy. And when they were come into the house, they saw the young child with Mary his mother, and fell down, and worshipped him: and when they had opened their treasures, they presented unto him gifts; gold, and frankincense, and myrrh.
—*Matthew 2:10–11*

According to the custom of the priest's office, his lot was to burn incense when he went into the temple of the Lord.
—*Luke 1:9*

Frankincense is not essentially an herb, but it is so generally known from Scriptures that it deserves a place in this book. Actually, frankincense is a resin produced by certain trees that grow in the Middle East, especially in the dry country in southern Arabia and northern Africa. It is white or colorless, produced by several species of Boswellia, the main source being *Boswellia sacra*, which is a shrub or small tree growing on both sides of the Red Sea. The tree resin is obtained by slitting the branches, which makes them "bleed" resin, and then collecting the exuding sap, sometimes called "tears" of resin. This is then burned as incense in religious rites or as a personal fumigant. In the Bible, frankincense was prescribed for a holy incense mixture. Other Biblical and botanical researchers have further described how frankincense is obtained for use today. First, they explain, a deep, longitudinal cut is made in the trunk of the tree. Below it the harvesters peel off a 5- to 6-inch narrow strip of the bark. That allows the milk-like juice to escape and be hardened by exposure to air. In several months the resin usually hardens into yellowish "tears" or globules, which are scraped off into baskets. Observers explain that the harvesting continues from May to September. That's when rains arrive to end the harvest.

History reveals that Herodotus had written that frankincense, in the amount of 1,000 talents weight, was offered every year, during the feast of Bel, on the great altar of his temple in Babylon. The religious use of incense supposedly was common in ancient Persia, today called "Iran," as well as in Babylon. Another interesting historic sidelight is that the Romans were said to have used frankincense in religious ceremonials, on state occasions, and in domestic life, too.

Burning incense has been a traditional ritual for many different religions through the years. The Bible produces dozens of references to the use of incense. Frankincense was highly prized among scents by the ancients in the Holy Land. In fact, along with myrrh, it appears in one of the most often quoted passages in the Bible.

The story of the Magi, the Wise Men who came to pay homage

to the Christ Child, is retold during the Christmas and Epiphany seasons every year. That Scriptural quotation from. St. Matthew is included at the beginning of this section. Judging from the choices made by the Magi, frankincense and myrrh were considered to be in the same highly prized category as gold.

The frankincense trees, of the genus Boswellia, grow in southern Arabia, across Abysinia, and along the east coast of Africa. Other species are found in India and the East Indies. This tree is seldom available for growth outside of these regions.

Some authorities believe that *Boswellia sacra* is the tree that is mentioned in the Bible with regard to frankincense. Others suggest *B. carterii* as the true frankincense plant. Some feel that two other species can be considered, *Boswellia thurifera* and *B. papyrifera*. It is true that they all provide a similar gum that does burn readily, spreading a pungent perfume like the ancients did so frequently in their religious ceremonies. Frankincense literally translates as "free-burning," which indeed it is.

These related trees are usually of relatively large size and are related to the terebinth tree and to those scrubbier trees that yield myrrh. Generally speaking, the tree's flowers are star shaped and white with some rose tinting. Leaves are compound and include 6 to 9 leaflets. In some respects, these trees look like a mountain ash, which is popular as a specimen landscape tree in the northern United States.

Frankincense is mentioned more than twenty times in the Bible. Most often it is referred to for its use in religious worship and for its worth in bestowing honor or tribute. Kings and Chronicles, Nehemiah and the Song of Solomon, and Isaiah and Jeremiah reference many ceremonies in which this incense was burned. In olden times it also was used for fumigating and purifying as well as burning in the temples.

Some Biblical gardens today are trying to obtain close specimens of Boswellia trees. In your garden you may elect to grow the mountain ash, which is similar in appearance but not actually related. Or, you may believe that the closest you can get is the tree that produces the common European frankincense. That is reportedly obtained as resin from slits in the bark of the

Norway spruce fir, *Abies excelsa*. But again, it is not really related to the true frankincense-producing trees of the Middle East area.

HYSSOP: *HYSSOPUS OFFICINALIS*

> And ye shall take a bunch of hyssop, and dip it in the blood that is in the bason, and strike the lintel and the two side posts with the blood that is in the bason; and none of you shall go out at the door of his house until morning. For the Lord will pass through to smite the Egyptians; and when he seeth the blood upon the lintel and on the two side posts, the Lord will pass over the door, and will not suffer the destroyer to come in unto your houses to smite you.
>
> —*Exodus 12:22*

> Then shall the priest command to take for him that is to be cleansed two birds alive and clean, and cedar wood and scarlet, and hyssop.
>
> —*Leviticus 14:4*

> And the priest shall take cedar wood, and hyssop, and scarlet, and cast it into the midst of the burning of the heifer.
>
> —*Numbers 19:6*

> And he spake of trees, from the cedar tree that is in Lebanon even unto the hyssop that springeth out of the wall: he spake also of beasts, and of fowl, and of creeping things, and of fishes. And there came of all people to hear the wisdom of Solomon, from all kings of the earth, which had heard of his wisdom.
>
> —*1 Kings 4:33–34*

This lovely purple-spiked, fragrant, culinary, and medicinal herb mentioned above so frequently deserves wider recognition. Hyssop has been used for ritual cleansing as noted in Leviticus (Leviticus 14:4, 49) and sprinkling of blood in the tabernacle

(Exodus 12:22), but some authorities believe these references are to the white marjoram, *Origanum syriacum* or *Majorana syriacu,* which grows commonly in rocky places and is related to mint.

In Psalms 51:7 David's confession of sin and humble prayer for forgiveness tells again of the hyssop: "Purge me with hyssop, and I shall be clean: wash me, and I shall be whiter than snow." If you also read again the passages of Numbers, hyssop appears in 19:6: "And the priest shall take cedar wood, and hyssop, and scarlet, and cast it into the midst of the burning of the heifer." Later, in verse 18: "And a clean person shall take hyssop, and dip it in the water, and sprinkle it upon the tend, and upon all vessels, and upon the persons that were there, and upon him that touched a bone or one slain, or one dead, or a grave."

It would appear, considering the different references to hyssop, that it must indeed be an herb that is valued for its particular properties in the rituals and in the cleansing of the people. Many writers and scholars debate which plant named "hyssop" the Bible is actually talking about, however. Some still argue that it is really the well-known garden herb we regularly call hyssop, *Hyssopus officinalis.* Others argue in favor of the caper, *Capparis sicula,* a spiny shrub found in desert areas and rocky parts of Israel. Still others favor the sorghum, *Sorghum vulgare,* as a likely plant. The word "hyssop" is unquestionably one of the most puzzling and controversial of all words in the Bible applying, or believed to apply, to plants. One fact is clear. The hyssop of our gardens today is not native to either the Holy Land or Egypt. In fact it is indigenous only to southern Europe, so it does not really fit into the proper understanding of what is truly a plant of the Scriptures and Holy Land. Because of the debates, however, I have tried to be open-minded about what I would include in a Biblical garden; therefore hyssop, whether it is a true biblical plant or not, is listed here.

Hyssop is an herb that is aromatic and slightly bitter and is found throughout the Mediterranean area. The plant name is probably derived from the Hebrew word *esob,* although other linguists believe that *esob* most probably referred to a local variety of marjoram. Remember that early translators of the Bible were at a disadvantage because there was no botanical nomenclature for them; they used their best guesses in naming flowers and herbs and other plants from the original Hebrew and Greek texts of the Bible.

Time-honored tradition identifies the hyssop of Scripture with the familiar herb Marjoram (origanum). There are perhaps six species found in the Holy Land of which the common kind, *O. vulgare,* grows only in the north. A related species, *O. maru,* thrives through the central hills and one variety is common in the southern desert.

Hyssop, nevertheless, is an attractive garden plant with dark blue flowers, but probably has only small value as a spice, because the aroma is weak and is reduced to nil after drying. Also, the taste is rather bitter. Herbalists advise, however, that hyssop can be used for dishes like potato or bean soup. Others suggest using it to spice chicken as an alternative to sage, which hyssop resembles in its slight bitterness but not in aroma.

Actually of Greek origin, the term "hyssop" comes from *hussopos.* Supposedly the term references the Hyssopos of Dioscorides, named from azob, a holy herb that was said to be used for cleaning sacred places. That thought is reinforced in the Scriptural passage, "Purge me with hyssop, and I shall be clean" from Psalms 51:7.

In the seventeenth century, British herbalist Nicholas Culpeper warned against taking it unless under the care of an alchemist, however. Nevertheless, written records note that during both the seventeenth and eighteenth centuries tinctures and teas were made and used by people to cure dropsy and jaundice. Hyssop tea also had been recommended for sore throats in old herbal books and a poultice of ground leaves was considered useful to promote healing of wounds and bruises. Such were the days of herbal medicine, before modern pharmacology arrived.

Actually, hyssop has a strongly medicinal smell that perhaps lured people to its use as a curative in the past. It has a camphor-like odor and does have some history of use as a cleaner. It also was utilized according to ancient texts to improve the smell of sick rooms and cooking areas. Today it has found other uses. Hyssop's minty leaves and flowers are used to flavor salads, soups, and stuffings. Leaves and flowers are used by herbalists for teas. The oil of hyssop has some use in perfumes according to botanists, and perhaps one of the more interesting sidelights is that the volatile oil from hyssop helps make some liquors including Benedictine and Chartreuse.

Beyond its Scriptural roots, hyssop has value for all gardeners. One delight is that bees, butterflies, and even hummingbirds are attracted to this plant. That helps around the entire garden as pollinators work to keep all plants well pollinated in blooming season. Although a member of the mint family, hyssop does not spread as intrusively as most mints so it can be mixed into landscape designs, beds, and borders for is appealing display. Organic gardeners have reported that hyssop repels flea beetles and other pests. In the past, bee keepers reportedly rubbed hyssop leaves on hives so bees stayed calm.

Plant Profile

Hyssop, an attractive, compact perennial member of the mint family is very aromatic. It stands upright as a bushy shrub with many branched, squarish stems and exhumes a typical mint-like smell when crushed. Flowers grow in whorls and form dense spikes at the top of main stems. The blooms are blue or violet, have four stamens, and are tubular with bell-shaped calyxes. Leaves are opposite, hairless, and about 1 ½ inches long. The plants usually grow 2 to 3 feet tall. Native to Asia and Europe, the hyssop is now naturalized throughout North America. Actually there are three varieties, known respectively by their blue, red, and white flowers, which bloom from June to October.

Growing Tips Hyssop is easy to grow. Start plants from seeds, cuttings, or by dividing parent plants. Pick a sunny spot with well-drained soil in full or partial sun. Sow seeds ¼ inch deep in rows 12 inches apart. Thin seedlings to 10 inches apart. Plant cuttings or divide plants in spring or fall. Hyssop needs only moderate fertilizing and can prosper in drier areas. Pruning is useful to keep these bushy plants under control. Clip periodically to 12 inches and plants will respond with tender new growth for use. Remove spent flower heads. Hyssop performs well as a perennial for 5 or so years. Then divide and replant to ensure tender new growth. This plant seems to repel insects, so organic gardeners suggest interplanting in veggie gardens may be useful. Because it is attractive and not intrusive, consider hyssop for planting in flower areas and vegetable beds, too.

Hyssop is useful as a kitchen herb for broths, soups, stews, and occasionally for salad. Hang bunches upside down in a warm, dark place. Then chop and store dried leaves, green stems, and flowers in tightly covered glass containers. Use only tender parts and avoid woody stems.

MARJORAM: *ORIGANUM SYRAICUM* OR *O. MARU L.*

> Then shall the priest command to take for him that is to be cleansed two birds alive and clean, and cedar wood and scarlet, and hyssop.
> —*Leviticus 14:4*

> And he spake of trees, from the cedar tree that is in Lebanon even unto the hyssop that springeth out of the wall: he spake also of beasts, and of fowl, and of creeping things, and of fishes.
> —*1 Kings 4:33*

Debate continues about the true herb in this and other Biblical passages. Because I have included hyssop in this early chapter, I want to include marjoram as well. Many Biblical and botanical scholars alike still seem unsure which herb to identify, and because there has been no unanimous decision on this issue for centuries, and probably won't be, I have included both hyssop and marjoram in this book as probable Biblical herbs.

Even today, travelers can find the bushy hyssop sprouting out of the ancient walls in the Holy Land. It is true, however, that many other plants do "spring out" of holes and crevices of walls in Israel as well. However, botanists who have focused their attention on this subject usually agree that the hyssop mentioned in the Bible is most likely the Syrian marjoram, as popular an herb today as it was in Biblical times.

Origanum syriacum is common among rocks and terrace walls, which fits the Scriptural passage perfectly. I vote with the majority on this opinion. Legends say that the Greeks called marjoram the "joy of the mountains" and used it medicinally. Other folklore reveals that garlands and wreaths were made of marjoram and used at weddings and funerals in ancient times. Myth tells us that this herb was precious to Aphrodite, the Greek goddess of love, because it was a gentle herb. In a romantic vein, it is said that if a person rubbed marjoram on their bodies they would dream of a future spouse. It is certainly true, also, that for years this marvelous, mild herb has been associated with marital happiness. A native plant to North Africa and southwest Asia, marjoram has naturalized throughout the Mediterranean area and is widely cultivated in North America today with good reason.

Marjorams are members of the mint family. Under favorable conditions they will mature to 3 feet tall. If restricted in rock crevices or poor soils, they usually are much smaller. Marjoram has erect, stiff hair branches with thick textured leaves. It typically bears white flowers. Examination of the hair-stemmed marjoram reveals that it is logical that it would hold water to serve as a sprinkler in religious ceremonies, as in Exodus 12:22: "And ye shall take a bunch of hyssop, and dip it in the blood that is in the bason, and strike the lintel and the two side posts with the blood that is in the bason . . ."

From its roots in western Asia and the Mediterranean, marjoram has spread its sweet scent to many lands. Both Greeks and Romans prized the aroma of marjoram and used sweet marjoram in creating crowns worn by bridal couples. As a flavoring, this herb has wider applications than most herbs. It is a natural with lamb and fish, but also adds its marvelous bouquet to eggs, vegetables, soups, stews, and stuffing for poultry, turkey, and game birds. Gourmet cooks point out that it is best to add marjoram to vegetables and soups just before the end of cooking time so its delicate perfume is not lost by overcooking.

Plant Profile

Marjoram is a tender, sweet-smelling herb that thrives as a perennial in southern areas but must be replanted as an annual each year in colder northern climates. It has a dense but shallow root system with an unusually bushy growth habit. Square stems branch frequently and are covered with tiny hairs. Flowers are tiny, white or pinkish, in spherical clustered spikes having 3 to 5 blooms. Leaves are ovate, fuzzy, and opposite. They are a pale gray-green color from ¼ to 1 inch long. Fruit, the seeds, are small nutlets colored light brown. The plant matures about 1 foot tall. Many gardeners note the similarity in smell to the botanically related oregano.

Marjoram likes warm weather, which enhances its fragrance. Sweet marjoram is a subtle, mild herb that begins to flower in July when it should be cut for use. It obtains its name of knotted marjoram because the flowers have roundish close heads like knots. Other choices for your garden could be similar wild marjoram, *O. vulgare,* or pot marjoram, *O. onites.* Sweet marjoram is usually preferred by most herbalists and in my opinion fits best as a Biblical plant.

Growing Tips Grow marjoram from seeds or cuttings in a sunny location. It prefers light, dry, and well-drained soil with full sun to produce its distinctive aroma and flavor. Plant seeds ¼ inch deep in late spring when soil is well warmed. Be very patient. Seeds may require several weeks to sprout. For this reason indoor planting in peat pots with soil mix is recommended. Keep moist until they sprout or they may die from lack of moisture. Then thin seedlings to leave the strongest two in each pot.

Be aware that marjoram is one of the few herbs that prefer sweet, slightly alkaline soil rather than neutral or acid soil. Because of this, add a little lime, a handful around each plant in the ground during spring and fall. Because seedlings are small it is best to weed by hand at first to avoid disturbing roots. Then hoe regularly or mulch to thwart weed competition. Water weekly, weed regularly, and enjoy ample supplies of this delightful herb. To keep plants from spreading, which they tend to do, prune every few weeks and remove blossoms. That also helps keep herbs sweeter. As plants mature in southern areas, divide them to share with others or plant elsewhere. Colder northern horticultural zones necessitate the need to replant each year from seeds.

Enjoy marjoram in salads, casseroles, or in making herbal tea. It also is a fine herb for seasoning meats, soups, and stews. Chefs rate marjoram especially useful for flavoring such heavy vegetables as legumes or cabbage. Potatoes spiced with marjoram are delicious. To save it, cut and hang branches in a dry, warm room that is well ventilated. A dark room will help preserve leaf color. When leaves are crisp, strip them from the stems and put away either whole or chopped in airtight jars.

MYRRH: *COMMIPHORA MYRRHA*

> Now when every maid's turn was come to go into kind Ahasuerus, after that she had been twelve months, according to the manner of the women, for so were the days of their purifications accomplished, to wit, six months with oil of myrrh, and six months with sweet odours, and with other things for the purifying of the women . . .
> —*Esther 2:12*

> All thy garment smell of myrrh, and the aloes, and cassia, out of the ivory palaces, whereby they have made thee glad.
> —*Psalms 45:8*

> Until the day break, and the shadows flee away, I will get me to the mountain of myrrh, and to the hill of frankincense.
> —*Song of Solomon 4:6*

> And when they were come into the house, they saw the young child with Mary his mother, and fell down, and worshipped him: and when they had opened their treasures, they presented unto him gifts; gold, and frankincense, and myrrh.
>
> —Matthew 2:11

Among all the passages from the Bible, the one from Matthew may be the best known and the one most quoted, especially during the Christmas season worldwide. As the Wise Men came to see the Holy Child, they brought with them gifts of what were their most valued treasures. What may not be so well understood is that two of these treasures offered by the Wise Men are actually derived from plants. As previously stated in this chapter, frankincense, which had a long history of use in religious ceremonies, is an incense that is obtained from trees.

Myrrh is a reddish-colored resin obtained from the myrrh plant, a spiny shrub, *Commiphora myrrha*, in a similar manner to frankincense. True myrrh was highly valued and esteemed by the ancients both as a perfume and as incense to be used in the temples during ceremonies. It also was used as an unguent and in embalming according to old texts.

Other texts note that this resin was sometimes dissolved in oil and either eaten with food, used as a medicine, or even cosmetically. It is sometimes difficult to verify old writings as authentic, but it seems worthwhile to gather such information together in one place when writing about individual herbs with such a unique heritage as myrrh. It is known that the aromatic resin of myrrh was valued at least 2,000 years before Christ. A Syrian legend passed along to the Greeks associates myrrh with the goddess Myrrha, daughter of Thesis who was a king of Syria. She escaped being killed by him, legend tells us, when the gods transformed her into a myrrh tree. From that intervention the legend continues that drops of gum resin that come from cuts on the tree are said to be Myrrha's tears.

Early in the Bible, Moses was instructed to annoint priests with an oil that contained myrrh. It is also known that the Egyptians used myrrh in embalming fluids and ancient texts say it was believed to be a cure for cancer and leprosy. According to other records the Egyptians used myrrh for fumigation because it burns slowly and reportedly kills certain insects, including mosquitoes. Whether these ages-old legends and folklore are true or not, myrrh continues be used in perfume, in making soap, and as an incense ingredient.

Because this desert tree is not grown in the United States, it may be nearly impossible to obtain because of plant import restrictions and ongoing problems in the Holy Land. However, myrrh may sometimes be available through some drug and health food stores as a tincture or even in gum form.

There is conjecture that the word "myrrh" in various translations of the Bible may also refer to a related plant, *Commiphora kataf*. Both of these trees are native to the coast of eastern Africa, Abyssinia, and Arabia. It is known from ancient records that the gummy substance taken from these trees provided the commercial myrrh of antiquity. Botanical texts and other references report that myrrh was widely used throughout the Holy Land and its adjacent areas by various peoples.

These two most likely species of myrrh are similar and forbidding in appearance. They are low, scrubby, and thorny shrubs or stiff-branched small trees. Today myrrh grow in rocky areas, especially in limestone hills in the Arab Middle East, Israel, and many parts of North Africa. One might assume that the trees are stunted—their knotted branches and support branchlets come out at right angles—but that is their usual growth pattern. The three-part leaves grow in clusters on the wood and are accompanied by sharp thorns. These minitrees or bushes do not grow more than 9 feet in height. Their leaves may be sparse, small, and very unequal. According to reports from the Holy Land, these shrubs were first recognized and identified about 1822 at Ghizan on the Red Sea coast. That district is so barren and dry that it is sometimes referred to as *tehama*, which translates to "hell."

Myrrh itself is actually a pale yellow liquid or resinous sap that oozes from fissures or wounds to the bark and hardens to reddish brown crystals. In commercial use these are usually found in "tears" of various sizes with the average being about walnut size. The surface is rough and usually brittle, semitransparent, and often has whitish marks. As incense, myrrh burns slowly.

Plant Profile

The myrrh plant is a small, thorny shrub that seldom reaches 9 feet tall. The leaves are divided into three leaflets with the one at the tip much larger than the pair that grows below. Leaflets are blunt, hairless, and roughly toothed. The shrub itself is scraggly looking.

Growing Tips Myrrh plants grow wild in their native habitat, but my research has not turned up any sources for these plants that would enable them to be imported into the United States. Equally important, duplicating its needed habitat would be virtually impossible.

SAGE: *SALVIA OFFICINALIS, SALVIA JUDAICA*

> And they shall eat the flesh in that night, roast with fire, and unleavened bread; and with bitter herbs they shall eat it.
> —*Exodus 12:8*

> And he made the candlestick of pure gold: of beaten work made he the candlestick; his shaft, and his branch, his bowls, his knops, and his flowers, were of the same: And six branches going out of the sides thereof; three branches of the candlestick out of the one side thereof, and three branches of the candlestick out of the other side thereof.
> —*Exodus 37:17–18*

All through the ages artists have borrowed from nature as they created their works of art. The Egyptians, Greeks, and Romans, as well as the Hebrews of ancient times, adorned buildings with replicas of plants and flowers. I could provide hundreds of examples, from the caps of pillars in Egyptian temples to the fleur-de-lis insignias used in battle flags and ceremonial clothing based on the iris flower. Suffice to say that artists have been captivated by plants since the world began.

Sage is not an herb mentioned by name in the Scriptures. However, it is doubtless that it is the herb referred to in the passage from Exodus that opens this section. According to Biblical scholars, however, this illuminating passage is referring only to the Judean sage, *Salvia judaica,* which grows up to 3 feet tall and can be found today in most of Israel.

Sage is a slightly bitter and aromatic herb and has a somewhat fruity fragrance. Today the main sage varieties used as spice come from the Mediterranean and Asia Minor. The Latin name *salvia* is from *salvere,* which means, basically, "save." That most likely traces to the early medicinal value of the plant. Although sage is an ancient spice, its use today seems to be concentrated on the Mediterranean from Spain to Grecian countries where sage-spiced foods are favored.

Herbalist records of the late 1500s mention different varieties of sage as a well-known herb in English gardens. Wild sage grows from Spain along the Mediterranean coast up to and including the east side of the Adriatic, where it can be seen in profusion on the hills in Croatia and Dalmatia. It seems to prosper near limestone with very little soil. It has been noted that the collection of sage forms an important cottage industry in Dalmatia. And some concur that the best sage grows on the islands of Veglia and Cherso; the surrounding district is known as the sage region.

In years long past sage was sometimes spoken of as *S. salvatrix,* which translates to "sage the saviour." Another old French verse salutes sage: "Sage helps the nerves and by its powerful might, palsy is cured and fever put to flight."

Through the centuries many kinds of sage have been used as tea substitutes. Old caravan records report that the Chinese preferred sage tea to their own and at one time in history offered three times the quantity of their choicest tea in exchange for that made with the herb. An interesting story but most likely herb folklore.

Varieties of *Salvia officinalis* are very diverse, including varieties with narrower leaves, red or variegated, and smaller or white flowers. The red sage and the white or green are classed merely as varieties of *Salvia officinalis,* not as separate species. *S. Lyrala* and *S. urticifolia* are well known in North America.

Note that sage is a very powerful spice and tends to dominate other tastes in cooking. Because it has a slightly bitter taste, it is also not appreciated by some people. The common sage, the familiar plant of the typical kitchen garden, is a shrubby evergreen that traces its natural habitat to the northern shores of the Mediterranean Sea. It has been cultivated for culinary and medicinal purposes for centuries in England, France, and Germany in Europe, as well as in North America, because it is hardy enough to stand most ordinary winters.

In a Biblical context, the Judean sage that grows to 3 feet tall has 4-angled stems that are stiff with paired leaves. When pressed flat, sage has been likened to the 7-branched candlestick that illuminated the tabernacle in the wilderness and later the temple in Jerusalem—the traditional Jewish symbol, the menorah. The sage plant, with its central spike and three pairs of lateral branches, bends upward and inward in a symmetrical pattern. On the branch are whorls of buds, which perhaps gave artists the "knops" on the Biblical golden candlestick.

Plant Profile

Sage generally grows about a foot or more high and has stems covered with downy hairs. Leaves are paired on the stem and are 1 ½ to 2 inches long, stalked and oblong. They are rounded at the ends and finely wrinkled. Leaves have a marked network of veins on both sides and are typically grayish green in color with small hairs. Flowers are borne in whorls and are purplish with lipped corollas. Plants typically blossom in August. All parts of the plant have a strong, scented aroma and a bitter taste. The fruit is an oval nutlet. The plant grows 12 to 36 inches tall, depending on growing conditions.

Growing Tips Sage is a hardy perennial that will thrive in full sun to light shade and prefers light, well-drained soil. The soil should be slightly alkaline, which can be accomplished by placing handfuls of limestone around plants each spring and fall. Keep soil moist until seeds or transplants are well started. Then, the plants need little moisture once they have become established.

Sow seeds in spring and transplant to 15 to 24 inches apart after seedlings are well rooted. Many herb gardeners prefer using cuttings or plant divisions because it may take 2 years to get mature plants from seeds. Most garden centers offer started plants these days. Divide old plants using the outer, younger plant areas. Dividing crowns every 3 to 4 years is a good idea to prevent plants from becoming too woody and less desirable. It is also possible to reproduce plants by layering. Simply tuck a stem in the ground, cover with soil, and hold in place with a stone or brick. When it sets its new roots, cut the sage apart from the parent plant. Pruning plants heavily in spring prevents them from setting seeds and promotes good leaf growth.

Use fresh sage leaves according to favorite recipes such as salads, or add to omelets, breads, and various meats and vegetables. However, most people prefer the stronger flavor imparted by dried sage. To dry sage leaves, snip from younger branches and discard stems to spread the leaves on clean sheets in the shade. Store when dry in an airtight colored glass or a solid container to preserve this herb's special flavor. Sage and onion dressing is an old British favorite.

WORMWOOD: *ARTEMESIA ABSINTHIUM*

> Least there should be among you man, or woman, or family, or tribe, whose heart turneth away this day from the Lord our God, to go and serve the gods of these nations; lest there should be among you a root that beareth gall and wormwood.
> —*Deuteronomy 29:18*

> My son, attend unto my wisdom, and bow thine ear to my understanding: That thou mayest regard discretion, and that thy lips may keep knowledge. For the lips of a strange woman drop as an honeycomb, and her mouth is smoother than oil: But her end is bitter as wormwood, sharp as a two-edged sword.
> —*Proverbs 5:1–4*

> He has filled me with bitterness, he has made me drunken with wormwood.
> Remembering mine affliction and my misery, the wormwood and the gall.
> —*Lamentations 3:15,19*

Wormwood definitely is not a pleasing herb as many historic references reveal. Other Biblical passages reveal that wormwood does seem to be held in low regard, for example, Proverbs 5:4: "But her end is bitter as wormwood, sharp as a two-edged sword."

The theme of bitterness and the ill omens associated with wormwood are even more clearly indicated in Lamentations 3:15, and later in verse 19, above. Then, in Jeremiah 9:15: "Therefore thus saith the Lord of hosts, the God of Israel; Behold, I will feed them, even this people, with wormwood, and give them water of gall to drink."

Other references to wormwood can be found in Amos and Revelations. It would seem, reading through these passages and contemplating their meanings, that wormwood was far from being a prized herb of Biblical days. Yet it was widely used. Consider these historic facts.

The earliest known description of wormwood was found in an Egyptian papyrus circa 1600 B.C. and authorities note that it was used to rid the body of worms. One legend about wormwood declares that it grew up in the trail left by the serpent as it slithered out of the Garden of Eden. Other old texts report that wormwood was long used as an insect repellent. According to another legend, the famed painter Vincent van Gogh had been imbibing absinthe, a powerful alcoholic drink made from wormwood, when he cut off his ear to give to a lady he admired.

Absinthe, an alcoholic drink that has led to serious mental problems and seizures, was made using leaves and flower tops of wormwood together with other aromatic plants. It is now illegal in many countries of the world.

Wormwood adapts well, however, to different climates. In some parts of America's "wild" West there are large tracts almost destitute of other vegetation besides wormwood or, as it is more generally known, sagebrush. It has been noted by ranchers that these plants are of no use as forage. The few wild animals that feed on them, if eaten, seem to have a bitter taste.

Plant Profile

Wormwood is a hardy perennial that seems unharmed by frost. It has a spreading habit and bears small, greenish yellow flowers in floret form on bushes that grow 2 to 2 ½ feet tall. Leaves are round to oval and are divided pinnately into finger-like segments. They are a gray-green color and covered with fine silky hairs. It flowers in July and August. Native to the Mediterranean area, this plant has naturalized throughout the world in temperate climates and is often cultivated.

The Latin name for wormwood is *artemisia* and Biblical scholars think that the wormwood in the Scriptures is most likely to be either *Artemisia herba alba* or *Artemisia judaica*. It is difficult to find seeds of these specific species in America but a number of artemisias, which are other types of wormwoods, are available in nurseries.

Growing Tips Grow wormwood in a well-drained, sunny area. It will do well even in poor soil. Plants grow 1 to 3 feet tall and have small leaves with somewhat hairy stems and yellow flower heads at the end of the branches. Seeds are tiny but germinate well and can be planted in fall to get a start or in spring. Space 15 to 20 inches apart and allow 3 feet for spreading of a mature plant. For additional plants, use cuttings or make plant divisions.

Be aware that wormwood has large amounts of absinthin, which is a substance that is toxic to other plants, and it will wash off leaves into the soil. Wormwood should have a special place separate from other herbs or valued plants in the garden. Although wormwood can be identified as a plant of the Scriptures, I do not recommend it for a Biblical garden on the grounds of its historic record and the problems noted.

By contrast, however, my good friend and advisor Reverend Marsh Hudson-Knapp, who tends a marvelous Biblical garden at his church in Fair Haven, Vermont, does grow several types of wormwood and reports that they provide a beautiful background and accent plant: mugwort, silver mound, southernwood, and fringed wormwood. As he explains, "They have thrived for years in our Biblical garden in close company with a wide variety of other herbs. Their silver foliage offers a nice contrast. They make a nice background in the garden and in cut arrangements and their meaning is a real part of faith. We do face bitter experiences and becoming bitter as wormwood is a temptation to Christians in times of trial. The plant offers a good reminder to us to forgive and learn to enjoy living in grace. I recommend growing it! However I do not recommend eating it!"

Chapter Three

Biblical Herbs: The Bitter Herbs and Food Herbs

> And they shall eat the flesh in that night, roast with fire, and unleavened bread; and with bitter herbs they shall eat it.
>
> —*Exodus 12:8*

> The fourteenth day of the second month at even they shall keep it, and eat it with unleavened bread and bitter herbs.
>
> —*Numbers 9:11*

Chapter 2 discussed all the major herbs that can be traced with accuracy to appropriate Scriptural passages. This chapter relates other herbs that scholars have researched and believe can be considered the Bible's bitter herbs.

One of the most knowledgeable scholars, Reverend Marsh Hudson-Knapp, says, "Every year our Jewish brothers and sisters celebrate deliverance by observing the Feast of Passover. One part of the seder meal involves eating bitter herbs, a reminder of how bitter life was for their ancestors when they were slaves." He continues, "Initially our ancestors felt thankful to be set free. In time, however, they began to grumble against God and Moses their leader. Their life in the wilderness had become difficult and discouraging. Many longed for the fruits and spices they had enjoyed back in Egypt. They remembered the fish we ate back in Egypt, the cucumbers... actually cantaloupe, the melons, the leeks, the onions, the garlic as cited in Numbers 11:5," he explains.

Yet, "eating with bitter herbs" despite its deep symbolism might also be considered one of the Bible's healthiest recommendations. Over the years, the bitter herbs have variously been interpreted to include chicory, dandelion, endive, lettuce, sheep sorrel, and watercress. Noted Israel botanical writer Michael Zohary adds dwarf chicory and poppy-leaved reichardia and also rocket. Others have added watercress as a potential contender to the list of being a bitter herb. Today, we know that they're good for you—and your garden.

CHICORY: *CICHORIUM INTYBUS*—ENDIVE AND ESCAROLE

Chicory has long roots indeed, both figuratively and literally. Not only is this a tap-rooted plant, those roots trace back to ancient Egyptian times more than 5,000 years ago. As I dug into the history of this herb, it was possible to find other interesting references besides the Bible. Chicory was grown along the Nile River and well known in Egypt among other places. There it was called succory and was eaten by Egyptians as a vegetable or in salads. It also was mentioned by Virgil, Ovid, and Pliny in ancient writings.

Dioscorides mentions chicory in his herbal writings in the first century A.D. Later, Emporer Charlemagne seems to have respected this herb and included it in herbs to be grown in his gardens, no doubt for use in meals for himself and at court.

British herbalist Nicholas Culpeper wrote about chicory in the mid-seventeenth century. Research at the Monticello garden website (www.monticello.org) reveals that President Thomas Jefferson grew chicory in his gardens there in the late 1700s, most likely as a forage crop, which chicory has been known for in various countries through the years.

In Europe, chicory has been grown not only as fodder for sheep and cattle but also as a salad and vegetable for people. In fact, in Belgium the young and tender roots are boiled and eaten with butter as are parsnips and turnips. It has often been mentioned as the source of beverages made from the long roots that are dried, roasted, and powdered. According to legend, Queen Elizabeth I of England liked to drink chicory broth.

Research into the notoriety of chicory as a coffee substitute reveals that some of the earliest food frauds concerned the use of ground-up chicory roots mixed with real coffee. Such tricks led to some of the earliest pure food laws that required honest labeling of coffee to specify the amount of chicory that was included in it.

In defense of the use of chicory, early growers extolled its virtue as having a mildly sedative effect that offset the stimulation of caffeine in coffee. As recently as World War II, chicory was again widely used in coffee mixtures. Even today throughout the southern United States, chicory in coffee is favored and served.

Basic chicory, however, has not gained much favor as a garden herb in the United States. It is considered by some gardeners to be more a blue-flowered weed of roadsides and not worthy of garden cultivation. With regard to appearance, this deeply rooted perennial has a bristly stem and bears rigid branches. Flowers are 1 to 1 ½ inches across, dandelion-like in shape but sky blue in color. Leaves are broadly oblong or somewhat lanceloate and hairy. Those at the bottom are larger than top leaves. Fruit—the seeds—are brownish oval and the plant grows 3 to 5 feet high. It typically blooms in the spring in the South and summer in northern areas.

To grow basic chicory, plant seeds in a sunny location with rich, deep soil that will accommodate its long roots. Keep seedlings well watered and remove weeds so the plant can set its roots well without competition.

The terms "chicory" and "endive" are frequently interchanged because Witloof chicory has been erroneously named French or

Belgian endive. Witloof chicory (aka Belgian endive), however, has the name *Cichorium intybus L. var. folosum,* whereas endive and escarole are *Cichorium endivia.*

Endive has curly, finely cut leaves. Some recommended varieties include green-curled Lorca, Ruffec, which resists cold and wet conditions, and Salad King. Others include large green-curled white-ribbed Frisan, a crispy green that is heat resistant for southern areas, and a "baby" endive, Tosca. Endive and escarole are both considered somewhat coarse in texture and have a strong flavor that some people say is bitter, thus qualifying them as "bitter herbs."

Among so-called escarole with typical broad, crumpled leaf types, a grower can try broad Batavian full-hearted Grosse Bouclee, which is slow bolting, and Salanca, which is cold tolerant, for late fall or early spring.

Witloof, the endive type, produces tall, leafy plants with a healthy, strong root system. The head-like clusters of blanched leave are 5 to 6 inches long, compact, and tasty but somewhat bitter. These strong-flavored leafy greens are commonly used in salads. Both are cool-season vegetables like lettuce and are best grown like head lettuce from indoors-started or purchased transplants in spring and direct seeding in fall.

Try growing Witloof in boxes filled with sand and peat moss in a warm, dark indoor spot during winter. Or try forcing Witloof chicory in pots or tubs to produce crisp, cream yellow, blanched "chicons" for salads. Witloof requires a fairly rich soil mix of humus or loam with sand and peat. Plant seeds the same as outdoors, keeping them well watered, and enjoy these bittersweet leaves for much of the year as an indoor plant.

Endive provides its slightly pungent but tasty salad greens in spring from early planting outdoors or as a fall crop from midsummer sowing. It grows well in gardens but hot weather contributes to bitterness. Seed packets provide detailed sowing and growing directions, which are similar to lettuce cultivation. To blanch heads, tie the outer leaves around the heads.

Gardeners suggest broad-leaved Batavian or escarole with its large and slightly twisted lettuce-like leaves are worth growing. These form around plants up to 16 inches across. The heart is very deep, well blanched, and creamy white.

If you wish to grow what some consider the traditional, almost wild-type chicory and experiment with drying roots as a coffee substitute, seeds are available from mail-order companies.

CUCUMBER: *CUCUMIS SATIVA*

> And the daughter of Zion is left as a cottage in a vineyard, as a lodge in a garden of cucumbers, as a besieged city.
>
> —Isaiah 1:8

Some scholars believe that the cucumbers of Biblical Egypt were most likely the snake- or muskmelon *Cucumis melo*. They've also reasoned that the melons mentioned in the Scriptures were the watermelon, *Citrullus lanatus*. Depending on one's interpretation of their interpretation, the scholars may have a valid point or not. That's one of the curious aspects of researching Biblical plants.

Others, by far more numerous, believe that cucumbers mentioned in the Scriptures do indeed mean cucumbers as known today. They were absolutely grown in Biblical times and in fact are still grown widely and eaten throughout the Middle East and Holy Land. Actually annual trailing vines, cucumbers have been grown in the Middle East for more than 3,000 years and are among the top ten most popular vegetables in the United States. They also are one of the ongoing popular veggies around much of the world. Another important point about them is that for eons cucumbers were one of the foods that could be easily preserved by pickling with herbs, using another Biblical plant, dill. I have, therefore, included it here for the reader's enjoyment with some facts and hands-on growing tips for the garden.

According to ancient writings, cucumbers were known to the Greeks and to the Romans, too. Famed historian Pliny noted that the Emperor Tiberius enjoyed cucumbers at his meals. As we discussed at the beginning of this chapter, the Israelites in the wilderness complained to Moses that they missed the luxuries they had in Egypt, actually naming cucumbers and melons. In the mid-eighteenth century it is noted by Hasselquist that these crops were food of the lower-class people in Egypt serving them "for meat and drink." Yet, although cucumbers have been a diet of the masses, the nutritional value is really negligible because about 95 percent of the composition of the fruit is actually water. Nevertheless, even today cucumbers are major crops and foods in Egypt along with other common foods of the masses such as onions, leeks, and melons.

Some botanical authorities believe that cucumbers originated in India. They also note that the cucumber, *Cucumis sativus,* is a member of the gourd family as are melons, squash, and pumpkins. Cucumbers grown for pickling and those grown for fresh market, commonly called "slicers," actually are the same species. Fruit of fresh-market cucumbers are longer and smooth rather than bumpy. They have a more uniform green color and a tougher, glossier skin than the fruit of pickling types. Leaves are large and form a canopy over the stems and fruits. Fruits are commonly harvested while still green and for small whole pickles when very immature. Happily, the more the cukes are picked the more the plants will produce.

Dozens of different varieties exist that have subtle differences in flavor for growing in a garden. It actually is a good idea to try several different types, especially those that are featured for their "burpless" qualities. Also new varieties defy such diseases as the wilts, which can kill nonresistant vines. Others, like new Bush Crop, produce prolifically on a compact bushy plant with very short vines. Even an apartment dweller can grow these in pots. The attractive, full-size fruit matures to 6 to 8 inches long and has fine tasting, crisp flesh.

Plant Profile

Cucumbers attach themselves to objects by means of tendrils. Leaves are mostly long-petioled and incised to make five points. Flowers are golden yellow and bell shaped. Some plants need room to roam because the vines sprawl. Newer varieties are space saving for smaller gardens but are very prolific. Check catalogs and descriptions on started plants or seed packages for growing details. Also look for the different types of flavors and uses to be sure to get the type of cucumber preferred. Try two varieties one year and compare with others next year for tastefulness. Some varieties are specifically bred for making pickles.

Growing Tips Cucumbers prefer a sunny location. Considering their native habitat in the hot weather of the Holy Land and the Middle East, that seems natural. They also prefer rich, well-fertilized soil to perform best, which should be deeply prepared because cucumbers put down deep roots. If the soil is poor, mix in compost, peat moss, and rotted manure. Improving soil helps it retain moisture, which is important for cucumbers because they are primarily water. Key to cucumber success is providing ample water especially when plants are setting their fruits. They, like melons, are thirsty plants, so if soil becomes dry they just won't prosper.

Plant cukes in late spring when soil is thoroughly warm. Start seeds in pots indoors 4 to 6 weeks before setting them outside to gain growing time. Cucumbers are especially sensitive to frost, so never plant them outdoors until all danger of frost is over. Remember that even large size cucumber seeds can rot in the ground if soil is too cold or wet. Inhabitants in northern areas, take heart. Plant breeders have produced short-season varieties with wonderful taste and texture, so the delights of cucumber eating can be enjoyed wherever one lives. Since cucumbers are prolific, a few plants will suffice all season. Hill culture is the best growing method. Space seeds 4 inches apart in a group of 4 to 6 seeds, cover with ½ inch of soil, and firm it over them. When seedlings are 2 to 3 inches high, thin to the 3 strongest in the hill.

If planting in rows for lots of plants, space them 12 to 15 inches apart. Apply several inches of compost around plants to smother weeds, absorb the sun's warmth, and help retain soil moisture.

Depending on the variety, cucumbers should be ready for the table in 60 to 75 days from planting. For most tender cukes pick when they are small, and pick daily. Don't use any that have begun to turn yellow because they are past their prime. Among good disease-resistant varieties try Spartan Valor hybrid, which is an All America selection with vigorous vines that produce prolifically. It is highly resistant to mosaic and scab disease, too. M&M hybrid also is disease resistant.

DANDELION: *TARAXACUM OFFICINALE*

Most Americans consider dandelions a noxious weed to be ruthlessly eliminated wherever it shows its beautiful golden blooms. However, the dandelion is not only a probable Biblical plant considered as one of the "bitter herbs," but for centuries around the world this herb has been valued for food. In fact, in many civilized countries dandelion leaves are prized for salads. The roots are dug, dried, and used for food or pulverized and used to make beverages; the blossoms are the base for making tasty dandelion wine. Perhaps Americans have been influenced by the power of chemical herbicide advertising and the enormous lawn care industry.

Some expert gardeners, reporting overseas via the Internet, offered accolades for dandelions as a wholesome, nutritious, worthwhile food. It also deserves a bit of honest history to put it in its proper perspective. I'm glad to do that because I love the golden glow of dandelions. Each spring I relish seeing thousands of the beautiful golden blooms grace my lawn. They appear quickly, bloom in profusion, and I simply then mow them down and am back to a basically green lawn. My love for dandelions has earned me the springtime nickname of "Dandelion Dan" Swenson. Television crews come to videotape, and the interviewers, being good sports, stay to taste dandelion pizzas made with English muffins, fresh dandelion leaves, pizza sauce, and cheese. Washed dandelion leaves for salad and dandelion tea for a beverage completes the meal.

History mentions dandelions in tenth-century medical journals. In the sixteenth century, British apothecaries called the drug *Herba taraxacon*. By the nineteenth century, dandelions were pot herbs grown in Europe and America. Linguists say that the name is derived from the Greek words *taraxos*, which means "disorder," and *akos*, which means "remedy" or "cure." Most authorities focus on the French phrase *dent de lion* or literally "lion's tooth," which describes the deeply serrated shape of dandelion leaves. Lion's tooth seems more believable.

According to agricultural authorities, dandelions are important to beekeepers because the plant flowers each spring even in cool weather and provides a small succession of bloom through autumn. Scientists also note that many other insects feed on the dandelion flower nectar. One study counted more than 80 insects that visited the blooms. Dandelion watchers note other curious facts. Flowers are actually very sensitive to weather conditions. In good weather all the parts are outstretched, but when rain threatens, the entire flower head closes up, which it also does against the dews at night. You can actually watch it. By about 5:00 P.M. the blooms prepare for sleep. By 7:00 A.M. or so they open again in the morning.

Children through the years have enjoyed huffing and puffing the fluffy dandelion blooms as they go to seed, giving the plant a nickname of "blowball." When all the seeds have been blown away the disc remains bare, white, and surrounded by the drooping remains of the bracts. That appearance gives the plant another nickname, "priest's crown," which supposedly was a common one in the Middle Ages when a priest's shorn head was a familiar sight.

Considering its popularity in Europe, especially in France, try this bitter herb—but never from lawn areas that have been treated with any kind of chemicals. Pick tender young leaves before the flowers form. When dandelions bloom the leaves are much more bitter. Add young leaves to spring salads in combination with other plants such as lettuce and chives. Dandelion leaves also make delicious sandwiches. Just lay some tender leaves between slices of bread and butter and sprinkle with salt and perhaps a dash of lemon or salad dressing. My wife and I also have boiled dandelion greens just as we do with spinach. In fact, a Maine firm sells canned dandelion greens as well as fiddlehead ferns. I still drink dandelion tea and each spring, as "Dandelion Dan," I don my Dandy Dandelion T-shirt. His slogan is simple and down to earth: "If you can't beat 'em, then eat 'em."

Other points of interest: In England there is a dandelion beer and my grandparents said that both beer and wine were made from dandelions. Neighbors in New Jersey send their children out to pick dandelion flowers. Then they pour a gallon of boiling water over a gallon of blooms and let the mixture ripen for several days. They didn't share the recipe but many cookbooks include one. Adding sugar, a bit of ginger, the rind of an orange, and some yeast seem to be the key ingredients. Dandelion wine makers say it takes several months to produce a good wine, which is similar to sherry.

Another European tradition is to make dandelion coffee from dried and ground-up roots. As with chicory, dandelion roots were used as a coffee substitute in Europe and in the United States when coffee was unavailable during World War II. With the new popularity of herbs and health foods, dandelion tea and coffee can be found at health food stores today.

The "Defender of Dandelions," Doctor Peter A. Gail, has spent many years promoting the worth and values of dandelions. He has conducted annual dandelion cookoffs, given prizes to winners, and published a fascinating collection of dandelion recipes in his *Great Dandelion Cookbook,* which includes 160 pages of fascinating recipes. The Rutgers University herb and natural food specialist also has published *Volunteer Vegetable Sampler,* which is a collection of recipes for backyard weeds with plentiful information about them. Check out some of his collected folklore and fun at his websites: www.goosefootacres.com or www.edibleweeds.com. Snailmail is Peter A. Gail, Ph.D., Director, Goosefoot Acres Center for Resourceful Living, P.O. Box

18016, Cleveland, OH 44118. Telephone: (216) 932-2145; e-mail is: petergail@aol.com.

Plant Profile

The "lowly" dandelion is a herbaceous perennial that develops from a long milky taproot about the thickness of the small finger. It is white inside, dark on the outside, and exudes milky sap when cut. It can grow a foot deep. Flowers are glorious golden yellow, 1 to 2 inches in diameter on smooth hollow stems that rise from the center of leaves which form a rosette. They open in morning and close like clockwork each evening. Leaves are 3 to 12 inches long and very jagged along the margin and mostly hairless and dark green. The seeds are formed on fluffy puffballs. As many as 200 tiny, narrow seeds with their parachute tufts can be carried by the wind over great distances. Plants grow 6 to 12 inches tall. Dandelions are found in temperate climate areas throughout the world.

Growing Tips Despite the fact that most Americans consider dandelions as a weed instead of an herb, some mail-order firms actually sell seeds of special varieties that have a less bitter taste. Consider ordering them for your Biblical garden. Just plant a few seeds and they will sprout, set their taproots, and reward you with golden blooms for years. Leave a few of the wild dandelions in a flower bed or border as a representative of the "bitter herbs." Don't have any? Simply borrow and plant a puffball of mature dandelion seeds from a neighbor's lawn!

GARLIC: *ALLIUM SATIVUM*

Garlic is an herb flavored with a variety of odd legends, folklore, and tall tales. In truth it has been a valued and respected herb from the beginning of recorded history and probably long before. Common garlic is a member of the same group of plants as the leek and onion. There are more than 60 different kinds of onion and garlic known to exist, so it is understandable that garlic should be identified from the Scriptures. The Hebrew Talmud offers many kinds of food that are to be regularly seasoned with garlic. It remains a favorite flavoring with the Jewish people today.

Originally native of southwest Siberia, garlic spread to southern Europe where it naturalized. Historic records show garlic as a favorite herb and food, growing wild in Sicily and other parts of Italy.

There are perhaps more superstitions and ancient folklore about this herb than most others. One story claims that Satan stepped out from the Garden of Eden after the fall of Man and garlic sprang up from the spot where he placed his left foot. Long ago garlic was thought to possess magical powers against evil. Read Greek poet Homer's *Odysseus* and you will note that garlic was used to keep the evil sorceress Circe from turning him into a swine. In fact, garlic looms large in Greek and Roman times. According to Pliny, garlic was used by the Egyptians during oath taking. Virgil indicates that garlic was eaten by the ancient Greeks and Romans.

Garlic was praised by many as a medicine. There is a long list of maladies for which garlic was considered beneficial. Many ancient tales equate it with strength, although humorists say that may be a result of the strong smell, which chases people away. Ancient records, nevertheless, state that Egyptian slaves ate garlic when they were building the pyramids. Supposedly Roman soldiers ate garlic to give them strength in battles, perhaps because it was the herb of Mars, the Roman god of war.

Herbal history finds garlic mentioned in English texts of plants from the tenth to fifteenth centuries. Linguists believe that the actual name for garlic is Anglo-Saxon in origin from *gar*, meaning a spear, and *lac*, which means a plant. This is referring to the shape of garlic's leaves, which are long and narrow. Upon reviewing other herbals, it is obvious that garlic or "garlichas" has been part of mankind's diet from the earliest times.

An Egyptian document circa 1550 B.C. lists garlic as a remedy for problems ranging from bites and worms to headaches. Pliny claimed it could cure many ailments. In the Orient, old texts relate that garlic had been favored for treating respiratory and heart conditions. The Chinese have long used garlic to treat worms and cardiac ailments. Before we laugh off what may sound like ancient folklore, it pays to rethink the fact that before modern pharmaceutical developments arrived, herbs were indeed mankind's first line of medicine.

Tens of thousands of pounds of garlic are currently grown commercially in the United States, mainly in California, Texas, and Louisiana. As with many garden plants, however, it can be easily grown and savored—this delicious, but rather strong herb—in one's own backyard herb garden.

Plant Profile

Garlic is an onion family member with a compound bulb that is made up of from 4 to 18 bulblets or actual cloves. These grow in a tan-colored papery sheath. Flowers are small, white to pinkish, and appear at the end of a stalk rising directly from the bulb and are grouped together in a globular umbel. Leaves are long, narrow, and flat, much like blades of grass. The garlic plant will grow about 2 feet tall. They flower in spring to summer. The bulb is the only part that is eaten. To avoid garlic breath, try chewing on a sprig of parsley or mint. Parsley tends to eliminate the odor while mint helps mask it.

Growing Tips To grow garlic, follow the cultural methods for improvising soil as done for onions. The soil may be sandy, loam, or clay but garlic does best in a rich, moist, sandy soil. These plants also prefer clean cultivation, so pay attention to weekly weeding to eliminate competition from unwanted weeds around the garlic bed. Grow a few in pots if preferred.

Garlic prefers rich, deep, moist, and well-drained soil in full sun but can do fairly well in partial shade. Start with garlic sets, which actually are the bulbs, from mail-order companies or garden centers. A pound of garlic sets will plant about a 20-foot row. Plant garlic sets in early spring for a fall harvest. Garlic is fairly cold tolerant, so plant several weeks before the last spring frost. Plant cloves 2 inches deep and about 6 inches apart. Full sun produces the biggest plants. When flower stalks appear in summer, cut them back so plants devote their growing strength to produce bigger bulbs. To enjoy their beauty, however, let a few plants bloom. Organic gardeners point out that the pest-preventing properties of garlic, an allium family member, make companion planting a worthwhile idea to chase aphids away and perhaps other pests, too. Therefore, interplant garlic with tomatoes, eggplants, and cabbage-type plants. For indoor growing, a few cloves are all that is needed to grow garlic in a window box or as potted herbs.

Harvest bulbs in fall when tops begin to turn brown. Break them down or stop watering to force plants to dry. Then carefully pull the plants the same way as onions and place in a cool dark place to dry. It is best to separate bulbs if laid flat on papers. Braid the stalks and hang them in your kitchen for decoration and use when a recipe calls for a clove or two.

LEEKS: *ALLIUM AMPELOPRASUM*

Leeks are more slender and cylindrical than their close relative, the onion. Their leaves are flat, somewhat broad, and more succulent than onion leaves, which are seldom eaten by people. By contrast, the leaves of the leek are what are most used today because there really is not much of a bulb to them at all. Their flavor is similar to an onion but more pungent. In the Middle East and Europe leeks are traditionally used for seasoning in stew, soup, and meat cooking. The leaves are especially enjoyed in salads and chopped for flavoring.

Leeks are most likely of Eastern origin because they were commonly cultivated in Egypt at the time of the pharaohs. It has been written that the Roman emperor Nero was fond of leeks and held them in high esteem, eating them every month. Legend relates that behind his back he was laughed at and called "the eater of leeks," which seemed to mean that he was eating the food of the lower class.

Some schools believe that the leeks mentioned in the Scriptures are truly leeks as we know them today, *Allium porrum*. Others favor the idea that the leeks of the Bible are more likely a leguminous plant known as "funugreek," with its Latin name *Trigonella foenum-graecum*. Both plants were known ages ago in Egypt and the land of the Bible; both were popular as food. But a compelling reason to identify the leek as the plant mentioned in the Scriptures is the simple fact that it was—and is—widely used both in Israel and around the world among the Jewish people, in their cooking, as a time-honored herb and vegetable.

In England today, leeks are used primarily in soups and stews. In French cookery, the blanched stems are employed widely. The herbs are relatively hardy and can be grown rather easily. In colder parts of the United States these plants should be dug and heeled-in during winter. Some gardeners prefer to winter them over in sand-filled pots in cool, dark basements.

Plant Profile

The leek is a member of the onion family but does not form a real bulb. The thick, fleshy stalk is about the same diameter at the base and resembles a large green onion without a bulb and has leaves that are flattened like those of garlic. The plants are biennial and flower the second year.

Growing Tips Leeks grow best in a cool to moderate climate in a well-aerated soil with both good drainage and good moisture retention capacity and a pH of 6.5 to 7. Deep tilling is helpful so that a longer shaft will develop. Plant seeds or transplants during late fall or early winter or set plants out in early spring. Plant seeds several inches apart and be patient. Leeks can take up to 120 days from seed planting to maturity, so using transplants from retailers or starting plants indoors saves outdoor growing time. Varieties have been developed for resistance to cold and also shorter season maturity. Look for King Richard, Primor, Goldina, or Tivi. Thin plants to stand 4 to 6 inches apart in rows about 2 feet apart. Leeks look like large scallions as they grow. The long thick necks should be blanched to improve their flavor. Simply hoe or till earth up around the base of growing plants when they are about the size of a pencil and the necks will blanch naturally and make the edible portion longer and whiter. Water weekly so leeks grow their tenderest best.

Harvest leeks when they are about 1 inch in diameter and free from blemishes. Tops should be healthy looking and free of discoloration. Leek lovers advise that these delicious plants will keep for several weeks if maintained at temperatures near 35 degrees Fahrenheit and a relative humidity near 90 percent. Other gardeners report that they have left the leeks in the ground in winter and dug as needed. For cold areas it is best to dig and try to store indoors in moist sand.

LETTUCE: *LACTUCA SATIVA*

I include lettuce as one of the "bitter herbs" mentioned in the Scriptures. For the purposes of this book, where a plant can be positively identified, at least from one or two translations or versions of the Bible, I've included the herbs.

Lettuce is a plant with a long history and appears to have originated in Asia Minor. There is ample evidence that forms of lettuce were used in Egypt circa 4500 B.C. It also has been written that the Romans grew types of lettuce resembling the present romaine cultivars. Other ancient writings tell us that it was known in China about the time of the fifth century.

The leaves of garden lettuce, *Lactuca sativa,* are often bitter when unbleached, a fact corroborated by home gardeners. Of course, modern plant breeders have improved varieties, which to a large degree have eliminated the bitterness of earlier types.

Lettuce is actually a cool weather crop. It thrives when the average daily temperature is between 60 and 70 degrees Fahrenheit and should be planted in early spring or late summer. At high temperatures growth is stunted, the leaves may indeed become bitter, and the seedstalk forms and elongates rapidly, called bolting. Some types and varieties of lettuce withstand heat better than others.

Several distinct types of lettuce exist. These include leaf or loose-leaf lettuce, cos, romaine, crisphead, and butterhead. Leaf lettuce is the most widely adapted type and produces crisp leaves loosely arranged on the stalk. It is the most widely planted salad vegetable. Cos or romaine lettuce forms an upright, elongated head and is a tasty addition to salads and sandwiches. Butterhead varieties are generally smaller, loose-heading types that have tender, soft leaves with a more delicate sweet flavor.

Loosehead and crisphead lettuce varieties include the iceberg types common at supermarkets. These are adapted to northern conditions and require the most care and timing because of heat sensitivity. That means plant them to mature before hot weather to have high-quality heads. It pays to start seedlings indoors for early planting and then sow another crop in summer to harvest in the cooler fall. Reliable leaf lettuce varieties include blackseeded Simpson, an early crisp and delicately flavored type; Greenhart, which matures in 45 days from seed; and Prizehead, which is crisp with excellent flavor. Grand Rapids and Oak Leaf also are favored. More colorful reddish leaf lettuce varieties include Red Fire with ruffled red edges, Red Sails, and Ruby, which is one of the darkest red lettuce types.

Heading or crisphead-type lettuce are also similar to what is found in supermarkets. Great Lakes is a reliable standard variety that holds up well in warm weather and adverse conditions, providing well-folded heads. Iceberg can stand some hot weather, too, but be advised to plant these early and then later for cool fall weather growing. Ithaca tolerates heat, resists bitterness, and is slow to bolt so it is worth growing, too.

Butterhead-type lettuce is not usually grown commercially because it doesn't ship well, but it is a tasty addition to any lettuce collection. For home gardens the tender leaves and succulent flavor make this type superb. Buttercrunch is smaller but has heavy, compact heads. Summer Bibb and Fordhook are other favored varieties. Others include Divina, Butter King, Dark Green Boston, Little Gem, and a red one, Sangria.

Cos or romaine lettuces include Cimmaron, a dark red leaf type; Green Towers, which is an early dark green with large leaves; and Paris Island, which is a longstanding variety.

When considering lettuce types, leaf, Cos, and Butterhead lettuce can be planted anytime in the spring when the soil is dry enough to rake the surface. Two or more successive plantings at 10 to 15 day intervals will yield an ongoing supply of lettuce. Even with heat-tolerant types, most lettuce does not withstand hot summer days. It is best to plant in spring and stop sowing at least a month before the really hot days of early summer begin. Plantings started in late summer mature during cool fall weather. Here's a handy tip from lettuce lovers. Heat-tolerant varieties, mainly loose-leaf types, may be grown in the shade of taller crops through most of the summer if extra care is given to watering well and regularly in a soil of sandy loam.

Plant Profiles

Growing Tips Cultivate outdoor soil 6 to 8 inches deep in cool weather. Build humus content year by year by adding compost, manure, and mulch of old leaves, which can be tilled or turned under to improve the soil. Plant seeds directly after danger of frost is past; however, veteran gardeners urge starting seedlings indoors for transplanting or purchasing prestarted lettuce from stores.

Head lettuce should be transplanted in most locations and requires more care than other types. Start transplants for a spring crop indoors or in a cold frame and harden transplants outdoors so that they become acclimated to the conditions under which they will be grown. Cos, butterhead, and leaf varieties also can be transplanted for earlier harvest. If sowing seeds, plant seeds ¼ to ½ inch deep, about 10 seeds per foot in single, double, or triple rows 12 to 18 inches apart. Then thin seedlings to 4 inches apart for leaf lettuce and 6 to 8 inches apart for cos or butterhead. (Of course, the surplus seedlings can be eaten.) For crisphead types, space seedlings 10 to 12 inches apart in the row. To ensure plenty of lettuce, consider sowing hot weather–resistant types of loosehead and bibb lettuce directly outdoors as early planted lettuce is removed.

Lettuce has shallow roots, so be careful when hoeing or cultivating. Be aware also that lettuce needs lots of water to be as tasty and succulent as it can be. It is best to use frequent light watering so leaves develop rapidly and produce highest quality lettuce. Organic mulches also can help moderate soil temperature and eventually improve soil conditions. Although mostly water, lettuce likes to eat well, so apply up to 4 pounds of 5-10-10 or similar fertilizer per 100 square feet of garden, or as recommended on the package for the type fertilizer preferred.

Cut leaf lettuce as soon as it is large enough to use. Cutting every other plant at ground level gives remaining plants more space for growth. Dark green lettuce leaves indicate higher fiber, flavor, and nutritional value. Although lettuce is not known for a high nutrient value, it does provide dietary fiber and vitamin A and potassium. The vitamin A comes from beta carotene and the darker green the more beta carotene. Another nutrition note: lettuce, except iceberg, is a moderately good source of vitamin C, calcium, iron, and copper according to food specialists.

One final point about lettuce and leafy greens. There has been a major move to mesclun mix greens. "Mesclun" is the term applied to a blend of fresh, tender greens combined for their variety of textures, flavors, and colors grown and marketed together. Leaves are harvested by cutting and plants are allowed to regrow. The actual ingredients vary, consisting of half a dozen or more of any of the following lettuces and other greens: looseleaf, red leaf, oakleaf, romaine, miners lettuce, and others mixed with arugula, chicory, dandelion, endive, radicchio, sorrel, spinach, parsley, watercress, and herbs such as basils and borage. This new phenomenon sprouted in France and has taken America by storm. Mesclun seed mixtures are available in garden center seed racks and especially in mail-order catalogs.

MUSKMELON: *CUCUMIS MELO;* WATERMELON: *CITRULLUS VULGARIS*

> We remember the fish, which we did eat in Egypt freely; the cucumbers, and the melons . . .
> —*Numbers 11:5*

Again we find debate among Biblical and botanical scholars, not about whether the melon actually is a Biblical plant, but about which melon the Bible is talking about. Since the common muskmelon has grown in the Holy Land since ancient times, as well as watermelons, it is included here.

During the awesome heat of the Holy Land while the tribes of Israel wandered in the desert wilderness, thoughts of the cooling refreshment of the melons that they had enjoyed in Egypt filled their minds. Even today, during scorching or steamy summer heat, melons have understandable appeal. Who today can resist the tantalizing aroma of ripe melons when summer heat sizzles?

Botanically speaking the muskmelon is thought to be a native of South Asia, from somewhere between the foot of the Himalayas to Cape Comorin where it grows wild. The cultivation of this melon in Asia traces back so far, in fact, that dates cannot be properly established. Botanical scholars report that the common melon, the muskmelon, was grown by the Egyptians. Other early records prove that both the Romans and Greeks were familiar with it.

For years in America muskmelons and cantaloupes have been considered the same melon, but technically that is not correct. Cantaloupes, *Cucumis cantalupensis*, authorities believe, had their origin in Persia and the nearby Caucasian region. Records indicate that they may have been first brought to Rome from Armenia in the sixteenth century. For all intents and purposes, however, I consider them the same Biblical herb, if obtaining one over the other is difficult.

As noted earlier, melons of various types are basic foods in Egypt and the Holy Land. The watermelon, *Citrullus vulgaris,* is thought to be a native of tropical Africa and the East Indies where it grows to giant size, 30 to 50 pounds. Watermelons traditionally have a smooth rind and generally are oblong and 1 to 2 feet long. These types of melons have been cultivated since earliest times in Egypt and were known in southern Europe and Asia before the Christian era. In Egypt the watermelon was used by the common people to calm fevers.

Although opinion is again divided, it is likely that the word "melon" translated from the early Hebrew word *avatiach* or *avatichim* really refers to both the muskmelon and watermelon.

The muskmelon is a tender annual trailing herb with palmately lobed leaves. Many people consider melons a fruit rather than a vegetable but scientifically they are classified as herbs—this is something the writers of the Bible also knew. As you read the Scriptures you may be surprised to find no reference to vegetables but many passages about herbs.

The watermelon varieties prized today for having fewer seeds may not find favor elsewhere. In Egypt, even today, watermelon seeds are valued. They are saved to be roasted, salted, and eaten as a popular dish, much like the sunflower seeds in America. As farmland has been restored to productivity in Israel, melons are a major crop. The revitalized agricultural areas between Jaffa and Haifa can produce melons of incredible size and delicious flavor.

Muskmelon Plant Profile

Muskmelons and the related cantaloupes are classed as tender annual trailing herbs. Flowers appear near maturity and bear the melons. Tendrils on the vines enable them to climb somewhat, but for most conditions they are best left to spread on the ground. Consult mail-order catalogs to find a surprisingly wide range of varieties, even space savers; all are related to the Biblical melons.

Typical melon flowers have bell-shaped corollas and are either male or female. Both kinds are borne on the one plant. Although this section focuses on two key melons, as gardeners and melon lovers know, there are many other types of melons that have a great diversity in the size and shape of the fruit. Melon flesh may be white, green, or orange when ripe, scented or scentless with varying sweetness.

Watermelon Plant Profile

Watermelons are annual trailing vines also and new varieties have been produced by plant breeders that require much less space for home gardeners. For those who live in northern climates, there are now short-season watermelons, which allows one to grow and savor their goodness. Starting melons in pots indoors offers even more growing time. The result is many more mature musk- and watermelons to enjoy, especially if they are planted in ground covered with black plastic to absorb heat and stop weed competition.

Growing Tips Enjoy both muskmelons and watermelons by using the short season–maturing varieties described above and those bred for bushy or compact plant growth that yield full-size melons. Muskmelons prefer warm, rich soil. Generous amounts of organic matter help improve their growth. Apply ample fertilizer to speed growth for tastiest melons. Read and follow directions for the type of plant nutrients purchased. I prefer to scatter some prilled or pelleted fertilizer around the melon hills after planting. Then later in the season, especially if I plant melons with black plastic mulch, I apply liquid fertilizer by hose end sprayer to nourish melons as they mature.

These Biblical melons can't stand cold weather, however. It is best to start a few seeds in peat pots indoors about 6 weeks before outdoor planting time. That gives a real jump on spring. If that is not desirable, plant directly in the ground, but wait until after the last spring frost. The hill system seems best. Prepare several hills by mounding soil slightly, leaving a central depression. Place 4 to 6 seeds in this 6-inch diameter and cover with soil, firming it well. When seedlings are 2 to 3 inches tall, mulch with compost, grass clippings, or other organic mulch to retain soil moisture and suppress weed competition. Actually I still prefer planting melons under black plastic mulch because it really stops weeds and absorbs the sun's heat to encourage early plant growth. Simply cut a cross slit in the plastic, plant seeds, and then cover the plastic with a light layer of grass clippings to make it more aesthetic.

All types of melons are thirsty plants, especially as the fruit forms and matures. That's why mulch is so helpful; it stops soil moisture evaporation and weeds that compete for water and nutrients. Besides, mulching saves weeding and hoeing time. Be sure to water vines each week and more during droughts as melons form and grow. That is vital to ensure proper size and more important, the necessary sweetness.

Yes, you can enjoy melons from container gardens. Because the growing area is restricted be sure to water several times a week and feed with liquid nutrients during the growing period as fertilizer package directions specify. For containers use only the bush-type muskmelons.

Thanks to plant breeders who have tapped the genetic pool of muskmelons dating back hundreds of years, there is a wonderfully wide selection of varieties. Here are some. Sweet'n Early hybrid is an early maturing treat with firm, bright, salmon-colored flesh that is sweet and juicy. Vines are resistant to powdery mildew and bear 6 to 8 melons per plant. Honey Rock and Samson hybrids combine disease resistance with heavy sets of fruits. Iroquois is one of the tastier home garden varieties with a small seed cavity. Hearts of Gold is another sweet one. Burpee Hybrid is larger. My favorite is Ambrosia with luscious, light salmon flesh that is firm, very sweet and juicy, and also has disease resistance. Today many veteran gardeners focus on varieties that are not just the sweetest and most productive, but also are disease resistant. It pays to take advantage of these new breeding developments to avoid use of excess chemicals to control plant diseases.

Watermelons, like muskmelons, prefer lots of sun and warmth. Like the muskmelon and cantaloupes, plant after all danger of frost in well-drained loam that has ample organic matter. Again, I prefer the hill system with 4 to 8 seeds per hill; thin to strongest 3 plants. Again, focus on disease-resistant new varieties. Some old tasty types are susceptible to wilt and anthracnose, so grow those that have built-in genetic resistance. With watermelons and muskmelons, it pays to grow different new varieties in comparison with old favorites. Marvelous, tasty new varieties are being introduced regularly, so watch for news in mail-order catalogs and local garden columns, and ask for updates at local garden centers.

One perennial question about growing melons is: When are melons ripe enough to pick? Some veteran gardeners say to tap the melon for a dull thumping sound. Others tug to test a melon's ripeness. Some new varieties can be identified when they are ready by coloration. However, over the years I've found that the most reliable way to tell when a melon ripe and ready for tasty eating is to give it a slight tug while it is on the vine. If it separates easily it is most likely ready to be picked, since it comes away in the hand. If it resists the pull, give it another few days to ripen fully.

ONION: *ALLIUM CEPA*

> We remember the fish we ate in Egypt at no cost...
> also the cucumbers, melons, leeks, onions and garlic.
> —*Numbers 11:5–6*

In my opinion, the most appropriate onions for a Bible garden are the Egyptian or Bermuda types. Both were and are eaten in Egypt today as vegetables.

Onions are often touted as a harsh or bitter vegetable. However, the onions so well known and widely grown in Egypt and other parts of the eastern Mediterranean and Holy Land are far different from the hard-coated types that come to mind. Classic Egyptian onions are, as are the popular Bermuda onions many gardeners know and grow today, sweet and mild. It is not exactly known when the onion came into being. Onions were grown in Ancient Egypt and eventually arrived in Rome. Their name, probably from the Latin word *unio,* means large pearl. The status of the onion in modern times rose after French onion soup was made popular by Stanislaus I, a legendary King of Poland.

Most researchers agree that the onion has been cultivated for 5,000 years or more. Onions may be one of the earliest cultivated crops. They were less perishable than other foods and were transportable and easy to grow in a variety of soils and climates. Though the place of the onion's origin remains a mystery, many Egyptian writings describe its importance as a food and its use in art, medicine, and even in mummification.

Onions grew in Chinese gardens more than 5,000 years ago. They also are mentioned in some of the oldest writings from India, historic accounts report that the Sumerians were growing onions as early as 2500 B.C. In Egypt, onions can be traced back to 3500 B.C. Onions were actually an object of worship because they symbolized eternity to the Egyptians who buried them along with their pharaohs. Because of its circle-within-a-circle structure, Egyptians apparently saw eternal life in the anatomy of the onion. Paintings of plants appear on the inner walls of the pyramids and in the tombs of both the Old Kingdom and the New Kingdom.

In the Great Pyramid of Cheops there is an inscription that tells of the price paid, many talens of silver, to buy onions and other food for the workers while the giant pyramid was being built. One scholar has estimated that at today's valuation more than $3 million dollars were spent for onions, garlic, and other root-type crops to feed the workers.

Today, thanks to demands of modern gardeners for tastier food plants, breeders have been developing and perfecting new, milder, and tastier types of onions. They have even caught the attention of the National Garden Bureau and other key American gardening groups.

Plant Profile

Despite differences in sizes and taste among different types, the onion is a leafy plant with a thickened base at its leaves, which is usually considered the edible part. The small, hard structure at the base of the bulb is the stem plate. Upper portions of leaves are cylindrical and hollow. The seed stalk that forms, usually during the second season, depending on the type of onion, may be 2 to 4 feet tall, topped by a globe-shaped flower. You may have seen this in gardens since allium is gaining favor in the form of flowering plants too. Some reach 3 feet tall with giant purple globe blooms that are surprisingly appealing and increasingly popular. Gardeners today have a wide choice of varieties and planting methods and can plant seeds, sets, or plants. Sets are the tiny bulblets grown by seed firms from seeds, then dried and sold for easier, faster, and more satisfactory results in most garden soils. However, even a wider variety is available from mail-order companies.

Growing Tips Onions prefer a very fertile and well-balanced soil with high organic matter from compost or turned under mulch and a sunny location. Spread lime if soil is too acidic. An onion's natural tendency is to grow its tops in cooler weather and produce bulbs in warm weather. They will first form a top and then when a specific combination of daylight, darkness, and temperature is reached, bulb formation begins. Plant onions in spring as early as the soil can be worked. It should be well warmed, in a sunny location, and have a crumbly texture. Place or push sets in 1-inch deep furrows and cover with soil. Use double rows if desired, staggering the sets in each at 3 inches apart. A pound of sets will give 50 feet of row space.

For something different try the French intensive system. This employs wide rows in which seeds or plants are placed more closely together, providing a much greater yield from each square yard. However, pay attention to adding extra nutrients and water. Other gardeners favor raised bed culture for onions. One advantage is such beds reduce bending and are easier to tend; they can also accommodate other herbs and vegetables.

A packet of onion seeds sows about 20 feet of row and offers a bonus—eating the seedlings that are thinned out according to label directions for spacing. Cover seeds with a fine soil or sand. No matter what planting method, keep weeds under control, especially when started from seeds. Clean cultivation seems to work better than mulching. Keep seedlings well watered.

Here are some favored onion varieties for the best eating so you, too, can capture the flavor of the ancients from Egypt and the Land of the Bible. Sweet Spanish is a very large, globe-shaped and fine-grained sweet onion. It has yellow skin with pure white, firm, crisp flesh. These mature 4 to 6 inches across. A white Spanish also is available with white skin but doesn't store as well as the yellow variety. Yellow Bermuda onions are extremely mild with juicy, white flesh. They are medium size, somewhat flattened onions. Yellow globe hybrid is an outstanding variety for early maturity. Crystal White Wax is another early, Bermuda-type onion, sweet and mild. Giant Red Hamburger is a hybrid onion that is tasty for slicing for sandwiches and salads. They have red skins with red and white flesh, colorful in salads. Ask around. As onions make a comeback in gardens many new varieties are now available.

Onions benefit from fertilizer, so application of 10-20-10 commercial fertilizer along the onion rows, about ½ cup of the fertilizer per 10 linear feet of row and covering fertilizer with 2 inches of soil seems to boost growth. Other types of fertilizer are okay, just be sure to follow the directions for the type you purchase.

When onion bulbs seem to have reached maturity, in 100 to 120 days from seed and shorter timing when using plants or sets, break off the tops, the hollow leaves. That hastens ripening of bulbs. Then, after a week or so, onions will ripen and can be pulled. Leave them on the ground in warm dry weather to dry. Or, bunch them together and hang them to dry on a wall or over a beam in the garage. Most onions store well when hung to dry.

Chapter Four

Flavor Your Life with Kitchen Herbal Gardens

Gardens during Biblical times, judging from Semitic literature and archaeological excavations were usually walled enclosures. In them paths led among herbs and fruit trees with canals of water, aromatic blooms, and arbors. Some of the more interesting readings are found in Genesis 2 and Genesis 3, Song of Solomon 4:12–16, Ezekiel 28:13, plus 36:35, and Joel 2:3.

Delving into other writings from ancient Babylonia, Assyria, and Egypt, scholars have discovered that the rulers of the time seemed to favor large gardens with a wide range of plants. Some drawings have been preserved that reveal park-like gardens with herb collections. It is natural that the large gardens would be favored, considering the Holy Land's two-season climate of arid, hot times and brief rainy growing periods. People needed shelters to escape the heat and gardens that were irrigated with precious water are proved landmarks of those times.

In fact, the inhabitants of the Middle East before and during Jesus' time were accomplished herbalists. From ancient writings we know that most people had plots specifically for growing herbs. Should you wish to explore some of the Biblical herbs in the Bible itself, refer to Hebrews 6:7, 1 Kings 21:2, Numbers 11:7–9, John 19:29–30, 1 Kings 4:33. Further readings lead you to the "bitter herbs" mentioned in Exodus 12:8 and Numbers 9:11.

Herb gardens, however, have been a valued asset through the ages. Throughout Europe and elsewhere herb gardens were faithfully cultivated within the cloistered walls of monasteries and also convents during the Middle Ages and the Renaissance. In addition to what the monks and sisters learned about herbs in the Bible, there were other important reasons for them to be grown by religious leaders and this was with regard to the healing arts. Monks dispensed a variety of medicinal herbs in an effort to alleviate the suffering and distress of their people. Actually, in those days, the pleasure of the herbs for their beauty and flavor alone was secondary in importance to their more pragmatic uses. There was no pharmacy industry so herbs were the sources of many treatments. Some are valid and have lasted to this day.

In addition to monks, friars, brothers, and nuns who grew herbs and administered to the people in the surrounding towns, like those in the Holy Land, the people of the Middle Ages and Renaissance themselves grew herbs. Again, they were used for their curative properties but they were also used for flavorings in cookery, as a source of color for dying fabrics, and in some cases for their fragrance alone. Herb gardens of the Middle Ages and Renaissance were usually laid out in simple, formal lines in patterned beds, with narrow paths for tending and harvesting and often with benches or places for meditation. During the Elizabethan era in England the herb garden was a treasured part of home and estate landscapes. Large decorative herb beds were plotted and grown and, ultimately, the American colonists naturally transplanted the herbs and ideas about them to the New World.

As we look to add herbs to our own gardens, one word of caution. Although herbs had been used for medicinal values in years past many uses were more folklore and legend than fact. Today with modern advances in medicine it is best to check with your doctor before even considering any herbal remedies. Old legends about herbs are not modern medicine so I do not recommend any herbal medication of any kind without consulting a doctor.

In terms of growing, the great majority of most commonly grown herbs prefer full sun and actually thrive best in poorer, more sandy or gravelly, well-drained soil. When grown in richer garden beds they make more luxuriant vegetative growth, but herbal experts believe that when grown in richer soil herbs tend to lose much of their fragrance and flavor. That may be more a matter of opinion. There are also a few herbs that prefer partial or light shade rather than full sun and a few which definitely prefer a moist location. You'll find details about each herb in the sections about each of them.

A quick overview about harvesting herbs is useful here. If the foliage is to be used either for fragrance or for flavoring, it is best to cut herbs just as the flowers are about to open. At this stage the essential oils are abundant. Actually, the best time to cut herbs is early morning after dew has dried but before the plants have been exposed to hot midday sun. When herbs are to be cut for the beauty of their flowers, such as lavender and chamomile, they should be cut in full bloom. However, if seeds are to be harvested and used, it is best to cut the seed heads when they are no longer green and the seeds have dried. Judging the right time may take experience because seeds such as dill can ripen and begin to fall rather quickly, but I'll discuss them as we go along. Plants grown for their roots should usually be harvested in the fall after plant growth has ceased.

This chapter discusses the following Biblical herbs: anise, balm, borage, caraway, chives, comfrey, fennel, horseradish, lavender, lovage, marrubium–horehound, mint, mustard, oregano, parsley, pot marigold, rosemary, rue, saffron crocus, and savory.

One last point before we delve into individual herbs. Because I'll be referring to historic herbalists periodically in this chapter and elsewhere, I wanted to put them and their expertise into proper context.

Pliny the Elder was a scientist and writer who had a passion for directly observing natural phenomena. He wrote a major book about natural history based on his observations and was especially interested in herbs. In addition to being a writer, Pliny was a Roman senator and the commander of the imperial fleet at the naval base of Misenum. His curiosity for natural phenomena seems to have been boundless. Once, when told about strange clouds on the horizon by his wife, he set sail to find out about them. It cost him his life because what he ventured into was the eruption of Mount Vesuvius in A.D. 79. A prolific writer, Pliny left 150 plus volumes of notes about many topics in addition to his respected *Natural History* volumes.

Dioscorides Pedanius of Anazarbos is usually simply referred to as Dioscorides. He was a Greek pharmacologist, if that is a term that can he applied to one of the ancients. A Greek contemporary of Pliny, his significant multivolume work, titled *De Materia Medica,* is thought to be the first systematic pharmacopoeia. It contains careful descriptions of approximately 600 plants and 1,000 different medications. This exceptional collection was translated and preserved by the Arabs and, ultimately, translated back into Latin by the tenth century.

Modern botanical scholars acknowledge that Dioscorides was the most authoritative writer in botany for sixteen centuries. Born in a small town near Tarsus, he was one of the most exceptional minds of the classical world of Greece and Rome during the time of Nero. As a surgeon with Nero's army, he traveled through Italy, Gaul, Spain, and even North Africa. This travel gave him access to all the major herbs of that civilized world. His herbal first published in 1516 has stood the test of approximately 1,500 years.

More recent herbalists include Nicholas Culpeper, who lived from 1616 to 1654 and also John Gerard, 1545 to 1607. Both English, their writings in the seventeenth century were the basis for modern herbalism.

Both were trained in medicine. Culpeper seems to have been particularly devoted to helping the common people. Culpeper, also spelled Culpepper, worked hard to make herb growing a fashionable part of the Elizabethan era. Many marvelous estates and great homes had glorious herb gardens attributed to him.

One of the most famed was Hatfield House in Hertfordshire where Queen Elizabeth I spent time in her youth. Other famed Elizabethan herb gardens were at Buckland Abby in Devon, which was the home of Sir Francis Drake, and another at Buckfast Abbey, an ancient monastery.

To review their work, see Culpeper's *English Physician and Complete Herbal* of 1651 and Gerard's *The Herball or General Historie of Plantes* of 1597, which may yet be available through some museums and major libraries. Some references will also be available on the Internet as herbal scholars input data.

With those bits of herbalist credentials in place for reference and research, I turn now to the task at hand, more of the amazing herbs of the Holy Land and the adjacent Mediterranean Sea countries.

ANISE: *PIMPINELLA ANISUM*

In ancient Rome, anise was used to pay taxes. By the sixteenth century it was supposedly being used as bait for mousetraps, if one believes that mice like the smell of licorice!

Pythagoras postulated that anise had medical values. Pliny stated that the herb was useful as a breath freshener. He also saw it as a benefit to youthful appearance. The Romans even used the herb for wedding cakes.

Delving into its Scriptural history, it is easy to assume that this dainty annual herb with its feathery leaflets was the plant mentioned in St. Matthew 23: "Ye pay tithe of mint, anise and cummin." My research, however, leads me to believe that it was not anise, but instead dill that was intended. But because there is a question about it, I have left it for discussion.

Being part British, I learned many little secrets about herbs from my grandmother, Amy Tudor Dugdale. Doing some genealogy, I was able to trace my Tudor roots back to 1500 and beyond into the *Domesday Book* that was commissioned by William the Conqueror. Even doing basic family tree research can lead to herbal information. Anise was mentioned in early English records in the 1300s and supposedly early British kings, including King Edward I, taxed the production of this herb. I even discovered notes about it in my great, great grandfathers confectionery business files, dug up by a cousin visiting Blackburn, England. It is truly amazing how herbs can have roots in the most incredible places.

Plant Profile

Anise is an annual herb that looks a lot like Queen Anne's lace. It grows with its feathery foliage about 2 to 3 feet tall. Leaves are on long stalks with lower leaves roundish, heart-shaped, and toothed. Upper leaves are more feathery and divided pinnately. Flowers bloom at the top of the stem in compound umbels and are small, yellowish white. Seeds—fruit if called by their proper name—are flattened, oval, gray-brown, about ⅛ inch long, and are somewhat downy with ribs. This herb blooms during the summer and is now naturalized throughout the United States.

Growing Tips To enjoy anise, sow the seed in dry, light soil, in a warm, sunny spot in April. Actually, plants will do well even in poor, light, and dry soil but prefer full sun. Sow seeds in rows 2 feet apart when soil is warm. Thin seedlings to 1 foot apart and keep weed-free. Moisten the seedbed until tiny plants get started.

Harvest seeds by cutting off seed heads when they are ripe. Snip into a bag to avoid seed loss, and dry on paper or sheets in sun or indoors. Then seal in glass containers for proper storage. Use the seeds whole or ground with cheese, carrots, in soups, stews, or as favorite recipes describe. Leaves also can be used in salads and as garnish for other foods.

BALM: *MELISSA OFFICINALIS*

A native of Southern Europe, especially in mountainous areas, and also with roots to the Holy Land, balm is called "sweet balm" or "lemon balm" as well. Linnaeus named it *melissa,* the Greek word for bee, probably because it attracts them. Pliny noted the attraction as well.

Balm is mentioned in Homer's *Odyssey*. The Greek physician Dioscorides noted that it was useful for scorpion and dog bites. He also favored putting lemon balm into wine for patients to drink; adding various herbs to wines was perhaps of little use except that the wine made patients feel better or at least more mellow.

More than 2,000 years ago Arabic writings, too, mentioned balm, where it was supposedly used for uplifting moods and helping with heart disorders.

British herbalist Nicholas Culpeper, who often became eloquent about herbs, wrote that balm "causeth the mind to become merry." Perhaps that thought can be traced to the fact that lemon balm, according to scientific research, is said to have a sedative effect on mice in lab tests. Whether the calming effect of balm is from a chemical property or a pleasing aroma is not documented.

Lemon balm has also been used to clean furniture, giving it a slight coat of oil as well as a pleasant aroma. Old agricultural texts suggested putting lemon balm inside new hives to attract new swarms of bees to settle there. There are also records of lemon balm being used in colonial American medicines and as a flavoring. Thomas Jefferson grew it among his many herbs at Monticello.

Plant Profile

Lemon balm is a loosely branching perennial that is somewhat scraggly in appearance but does offer an appealing lemony smell. Typical square stems identify it as a member of the mint family. Leaves are opposite, ovate, and toothed and can be up to 13 inches long. They are strongly lemon scented. Plants grow about 2 feet tall. Flowers appear in clusters at leaf axils, typical of mints, and are tubular, about ½ inch long. Plants flower July to September. This plant dies down in winter but since its root is perennial it reappears each year.

Growing Tips Balm grows well in any soil and can be propagated by seeds, cuttings, or division of roots in spring or fall as with other mints. Cut roots into small pieces, with several buds to each, and plant 2 feet apart in ordinary garden soil. Balms prefer full sun but will grow in shade, too. They like average, well-drained soil. When planting seeds, be patient because they are slow to germinate. Keep plants free from weeds, and in dry periods water them, but they need less than other mints. Plants take a year to root well and after that provide ample leaves. Harvest leaves before flowering for optimum fragrance. Cut off stems 3 to 4 inches from the ground and avoid bruising leaves. Dry in the shade on warm days on sheets or clean screens. Leaves are best for making tea and many herbalists use leaves chopped into salads or brushed on fish. Others suggest trying small amounts with vegetables to test taste appeal. Look up recipes in cookbooks and follow your taste buds as you cook. Add some lemon balm potpourris.

BORAGE: *BORAGO OFFICINALIS*

According to the Celts, this ancient herb allegedly gives people courage. Other writings from Roman times also allege that Roman soldiers took borage for the same reason and sixteenth century British herbalist John Gerard reminded us of the Roman saying *ego borago gaudia semper ago,* which means "borage brings me courage." Roman scholar Pliny noted that borage was said to lift the spirits. The renowned Greek writer Dioscorides indicated that borage "cheered the heart." Celtic folklore also said borage was the herb of gladness.

Botanical researchers report that borage probably originated in Asia Minor and North Africa and was found in the Holy Land. Today it grows in the Mediterranean countries, Europe, and North America.

Controversy remains over the source of the borage name. Some say the Latin *borago* is a corruption of corago, from *cor,* which means the heart, and *ago,* which translates to "I bring." Others linguists argue that there must be a connection between the plant's name and its hairy appearance and offer that a Latin term for wool, *burra,* and the modern Italian *borra* gave borage its name. Others focus on the Celtic warrior theme of courage and the word *barrach*, which loosely translates to a "man of courage."

In ancient times and in the Middle Ages, borage was known for its cooling quality and refreshing flavor as well as its ability to make men merry. Other historic texts relate that in medieval times, borage tea was given to competitors in tournaments as a moral booster. Some authorities note that borage also has been used to flavor wine—hard to imagine considering its cucumber-like taste.

Plant Profile

Borage is a self-seeding annual plant with hollow, bristly, branched, and spreading stems that grows up to 2 feet tall. Leaves also are bristly, oval, or oblong and lanceolate. The basal leaves form a rosette. Others grow alternately on the stem and branches. This plant tends to sprawl and become bushy into an attractive rounded shape. Older plants can become scraggly, so it helps to prune and use parts periodically. This flowering herb produces delightful blue or purplish, star-shaped flowers in loose racemes from June to August. That appealing blooming habit makes it useful for interplanting as well as in herb gardens. Bees seem to appreciate this flowering herb in season, too. Herbalists note that the star-shaped flowers have been used to decorate cold drinks, gelatin, and fruit salads. In the past, borage has been candied to decorate cakes and confections. Only the fresh flowers are used.

Growing Tips Borage is an easily grown annual but likes plenty of space in a sunny location. It is not a fussy plant and will thrive in poor but well-drained soil. However, be aware that the richer the soil the bushier the plant can become. It prefers full sun and protection from wind. Because it is somewhat brittle, borage can be blown over, so veteran herbalists advise placing several plants together to support one another, or staking plants.

Sow seeds when spring frost danger is over. Space seedlings 2 feet apart so the plant can attain its bushy growth. Keep weeded and mulched. Once growth is established borage plants tend to reseed. Transplant young plants to enjoy their blooms in other areas. Organic gardeners have noted that borage is an excellent companion plant for tomatoes and squash and seems to help nearby veggies resist insects and disease. However, no scientific research proves that point. An alternative is to grow borage in containers. Plant in fertile soil mix and keep well watered in a sunny location.

Borage flowers and leaves have been a traditional decoration for summer cocktails and other drinks. Other herbalists like the flowers and young leaves to garnish salads and dips to take advantage of its cucumber flavor. Fresh use is recommended because it does not dry or freeze well.

CARAWAY: *CARUM CARVI*

Most of us think of this as two words, caraway seeds. Today, caraway is used mainly in cooking and best known as an ingredient or topping for rye bread here in the United States. By digging deeper into the flavorful past of caraway, it traces its heritage back more than 5,000 years to seeds found in archaeological excavations.

Caraway is a member of the group of aromatic plants with carminative properties that include cumin, dill, and fennel. However, it is mainly used today for flavoring in cookery and baking. Caraway is a popular herb for flavoring meat in southern German and Austrian dishes as well as in vegetables. It has been popular in many German dishes from sauerkraut to Schweinsbraten. In Hungary, the famed Hungarian herdsman stew, goulash, is often created with caraway seeds. In Scandinavia, a black caraway bread has been popular for decades. Herbalists note that caraway's aroma does not blend well with most other spices, but it does combine well with garlic.

Oil from the seeds is used to flavor kummel, a liquor produced in Russia and Germany. The Swedes have their aquavit as a favorite beverage. Caraway leaves have also been used to flavor soups, stews, and salads and its roots are used as a vegetable in certain parts of the world, prepared like parsnips. Gourmet herbalists say that fresh leaves can be minced for cooking—in this regard it seems like the dandelion, used in the cuisines of so many countries but overlooked for its virtues in America where it is considered a weed. At least caraway has maintained its popularity for seeds as flavoring. Some scholars believe that caraway is an authentic herb mentioned in the Scriptures. However, many more experts believe that cumin is what is meant in those references.

Caraway has a rich tradition. Linguists say it originated with ancient Arabs who called the seeds *karawya*. Others note that the origin of our word "caraway" traces to the Latin *carvi*. Dioscorides recommended that caraway seed oil be taken "by palefaced girls." Other texts from the Middle Ages and in Elizabethan England refer to caraway. For example, in Shakespeare's Henry IV, Falstaff is offered "a pippin and a dish of caraways." According to sources in England, there remains a custom of serving roasted apples with caraway seeds at Trinity College just as it was during the reign of Elizabeth I. Reportedly, some old inns in Scotland offer caraway seeds during afternoon tea for dipping the buttered side of bread into for flavoring.

Using seeds over bread and cake has been a time-honored practice elsewhere as well. In Germany caraway seeds have been used in cheese and with soups. In Scandinavia, especially Norway and Sweden, caraway bread is fairly common and popular.

Plant Profile

Caraway is a biennial herb bearing a mound of finely cut, carrot-like leaves on furrowed stems growing 1 ½ to 2 feet high. It thrives in well-drained soil and bears umbrels of white flowers that blossom in June the second year. Some varieties do have an annual habit, flowering and going to seed their first year. The biennial varieties produce a long taproot that resembles a carrot or parsnip, although they are much smaller. As with many herbs, what are popularly called "seeds" are technically the fruits. The crescent-shaped seeds are oblong and pointed on either end with five distinct ridges.

Biennial caraway is the type grown commercially in Alberta, Canada, and is native to Europe and western Asia. In the seedling year, plants resemble carrots, growing about 8 inches tall with finely divided leaves and a long taproot. By the second year, 2- to 3-foot stalks develop, which are topped by umbels with pink or white flowers. These produce the seed, which is used as a spice.

Growing Tips Caraway thrives in well-drained soil in full sun to light shade and should be treated as a biennial. Plant caraway seeds ½ inch deep in rows 12 to 15 inches apart and cover seeds lightly with soil. Thin to 8 inches apart. Remove weeds or mulch well to grow plants best without weed competition. Water regularly as the plants set their taproots. Note that an autumn-sown crop will produce the following summer, ripening about August.

As crops mature watch carefully to harvest seeds before they drop. They are ripe when they turn brown. Snip stalks before seeds drop and tie in bundles, upside down in a warm, airy place. Seeds will drop onto a clean sheet or tray. Store seeds in an airtight glass jar. Favorite uses include making pickles, served with vegetables, and sprinkled on rye bread.

CHIVES: *ALLIUM SCHOENOPRASUM*

These tasty herbs have been pleasing mankind for more than 5,000 years. Native to the Orient, chives were well known and used and recorded in ancient Chinese writings. Following their favor in the Orient, chives found their way into gardens throughout the Mediterranean and Europe where they became a basic herb. Chives have been mentioned in Greek writings as an herb of good taste and worthiness. Today they grow wild in the Holy Land, but also are favorites in many herb and home food gardens there.

Chives did not seem to attract the fanciful attention of early writers like many other herbs did. However, some mentions allude to the magical powers of the plant without giving specific details other than that bunches of chives and onions were often hung in homes to ward off evil influences.

Chives are the smallest of the onion family. Relatives include garlic, leek, and shallot. Wild chives grow in rocky pastures throughout temperate climates. This tasty and versatile plant was known to the Greeks and to this day grows wild in Greece, Italy, and throughout the Mediterranean. Chives grow from Corsica and Greece to Sweden, in Siberia as far as Kamschatka, and also throughout North America.

Linguists explain that the Latin name of this species means "rush-leek" and it is indeed related to those two other plants. References to chives are found in Roman writings, but they seem to have been more an everyday accepted herb than one that was celebrated in any special way. Dodoens gave the French name for it in his day as *petit poureau,* relating to its rush-like appearance. In French today it is commonly called *Ail civitte.* French cooking combines chives with shallots, marjoram, tarragon, and other herbs. They nicely complement onions, potatoes, cauliflower, tomatoes, peas and carrots, and corn. Chives help flavor fish and poultry as well as cheese and egg dishes. Chives and sour cream on baked potatoes are a classic American dish.

Chives deserve a place in decorative gardens as well as herb spots. They are hardy perennials that provide a profusion of clover-like rose-purple blooms in the spring. They have multiplied into large clumps at our driveway gates, receiving rave reviews from guests who see them as they come to visit. Because these plants spread we keep several empty pots on hand. Whenever someone remarks on their beauty and then tastes chopped chives with their salad or soup we simply offer to pot a clump.

Chives are evergreen in mild areas but go dormant where winters are severe. Pot one or two smaller clumps for use during the winter as house herb plants. They will thrive on a windowsill along with parsley and provide tasty snippings whenever desired.

Plant Profile

As a member of the allium family, chives grow from tiny bulbs. The hollow green leaves shoot up and the plant forms a thicket of tasty leaves that can grow 18 inches tall. Individual leaves are dark green, slender, and cylindrical and taper to a point at the top, 6 to 10 inches long. Flowers are small and purple with dense globes at the top of stems usually in June. The seeds are formed in tiny capsules and are similar to onion seeds.

Growing Tips Chives are very easy to grow once they're started. Starting from seeds they germinate very slowly and require constant moisture and darkness. A better bet is to use divisions of plants or get pots of started chives at stores. Chives spread so fast that most gardeners must divide clumps every few years and are happy to give some divisions away. Plant clumps of 6 to 10 bulbs 5 to 8 inches apart in a sunny, well-drained spot. Providing regular moisture ensures quick rooting. As they grow, they'll drop their seeds and spread into larger clumps within a few years.

Snip chive "greens" from the clumps three or four times in the season. It is best to cut chives fairly close to the ground. New leaves will soon grow again and are more tender each time of cutting. Beside weeding between the clumps, they require little further care. Because they do spread, it is best to divide clumps or beds every few years. Use both the leaves and the flowers for flavoring—flowers are a welcome addition in making special herb vinegars. The flowers also are attractive in dried flower arrangements. Use chives additionally as companion plants to keep pests away from carrots, roses, and tomatoes.

For containers simply pot an extra clump of chives in late summer. Leave it outside until fall when leaves die back. Bulbs need a cold dormant time to send out leaves again. After they die down, bring the potted clump indoors to a sunny windowsill and leaves will sprout within a few weeks to enjoy all winter. When pruning for use, leave some leaves so the plant continues to grow. With a little bit of liquid fertilizer when watering, chives will grow well as an indoor container plant.

COMFREY: *SYMPHYTUM OFFICINALE*

Dioscorides prescribed comfrey to heal wounds and mend broken bones. Its generic name, *symphytum,* translates to "coming together," and Roman scientist and naturalist writer Pliny described how comfrey was so sticky that when meat chunks were cooked with it they became "glued" together. Other names for comfrey included knitbone, bruisewort, boneset, and gum plant. Old British records indicate that comfrey was regularly grown in gardens for its virtue in wound healing. In the Middle Ages it was a famous remedy for broken bones. It has also been cultivated by farmers as fodder for their livestock. During the Irish potato famine in the 1840s an Englishman envisioned that research into comfrey might help avert the blight. From this notion came the Henry Doubleday Association, now one of the largest organic gardening organizations in Europe (find their website at www.hdra.org.uk/).

Since the first record of its use in 400 B.C., comfrey has been claimed to be useful for treating burns, insect bites, and various skin problems. Others felt that it was a source of vitamin B12 and a protein builder. However, there are negative aspects. In 2001, the U.S. Food and Drug Administration requested that makers of dietary supplements containing the herb comfrey withdraw their products. Comfrey seems to contribute to liver damage and can possibly have a role as a cancer-causing agent. Comfrey preparations and drinks are also toxic and if taken internally act as a poison.

Plant Profile

Comfrey is a hardy, upright-growing perennial. It dies down in winter but comes back strong each spring. Its inch-thick rhizomes or roots can be up to a foot long, black on the outside and white on the inside, and contain a fleshy, juicy, and mucilaginous substance that gives this herb another name, slippery root. Leaves are lanceolate to ovate-lanceolate and join the stem at bases and can be up to 8 to 10 inches long. They are dark green and hairy and the plant can grow 3 to 5 feet tall. Stems are leafy, 2 to 3 feet high. Stout and hollow, they are winged at the top and covered with bristly hairs. Flowers are on short, curved racemes and are blue to whitish. Comfrey begins to bloom in April or May and usually continues all summer.

Growing Tips Comfrey does no harm just growing in the garden and it is worthy of a spot in Biblical gardens. It is an easy, fast-growing plant that does well as a background or foundation planting. It does tend to overgrow other plants so regular pruning is needed. Comfrey prefers moist rich soil but actually will thrive in almost any soil or situation. It likes full sun but does well in shady areas, too.

Propagate comfrey from seed by dividing roots in the fall or making cuttings at any time. Because it tends to grow large, space plants 2 ½ feet apart each way and keep well weeded until they have a strong roothold. Although attractive as an ornamental plant, comfrey is difficult to eradicate once established, because new plants pop up from severed portions of the root.

FENNEL: *FOENICULUM VULGARE*

Fennel was well known in early Greek history where its name was "marathron," derived from *maraino*, which means "to grow thin." The British herbalist scholar Culpeper concurred, saying that the fennel plant was used as a drink to help make fat people lean.

Ancient Romans praised fennel (the name derives from the Latin word *foenum*) for its aromatic fruits and tender, edible shoots. Naturalist scholar Pliny praised its medicinal properties and cited more than 20 remedies for it. The Anglo-Saxons also included the herb in medicinal mixtures and used it with their cooking (the herb has a taste similar to anise or licorice). Fennel shoots and seed were mentioned in ancient records of Spanish agriculture dating circa A.D. 960. Charlemagne encouraged fennel cultivation on imperial farms, and by the 1600s the plant was commonly eaten for its perceived value to aid digestion.

Elsewhere fennel is mentioned in *Paradise Lost,* where Milton writes, "A savoury odour blown / Grateful to appetite, more pleased my sense / Than smell of sweetest Fennel." Of more recent poetic time, fennel was praised by Longfellow, "Above the lower plants it towers / The fennel with its yellow flowers."

According to Biblical plant scholar F. Nigel Hepper, one of the unusual spices of the Old Testament was galbanum, which came from the large fennel, *Ferula galbaniflua*, although it is no longer available today. His recommendation for a fennel for a Biblical garden would be the common giant fennel, *F. cummunis* or culinary or sweet fennel, *F. vulgare*.

From the Middle Ages we learn that fennel should be hung over doors to keep out evil from the home. Other stories allege that fennel seeds should be eaten during lengthy church services to stop the sounds of growling stomachs.

Plant Profile

Fennel is a beautiful semihardy perennial plant. With this herb, seeds, leaves, and roots are used in different ways for different purposes. It has a thick taproot that resembles a long carrot. This produces branching, erect stems that grow 4 to 5 feet tall. These are smooth and typically blue-green color. Leaves alternately branch out from joints on the smooth stems and are pinnately divided into feathery segments. Flowers are bright yellow to golden in large, flat terminal umbels that may have from 12 to 20 rays. Bloom time is July to September. Because many people harvest fennel all season, it is rarely seen in its full and beautiful form. It can be a graceful ornamental.

Growing Tips Fortunately fennel thrives almost anywhere in average, well-drained soil and prefers full sun. Once established, plants will last for years. It easily grows from seeds sown early in April in rows 6 inches apart. Keep beds moist for the first few weeks until leaves appear. Then, cut back because fennel is adapted to dry and sunny situations. Note that fennel doesn't grow well with some plants. Gardeners point out that it can have a damaging effect on bush beans, dill, and tomatoes. Another point is to avoid planting fennel near coriander or wormwood since this can reduce the fennel seed set.

Several plants ensure plenty for harvesting leaves, as well as stems for salads and other recipes. Use fennel in vegetable dips and in cream sauces, even when grilling fish. A little goes a long way, however, and can be overpowering. Snip leaves once the plants are well established. To use stems, periodic pruning is required, so grow several plants. For seeds, as with other herbs, be alert as seed heads ripen, turning from yellowish green to brown. Check daily and cut the entire seed head into a paper bag and store in a warm, dark place. Once seeds are dried, shake and remove them and store in a tightly covered glass jar. If seeds are not wanted, cut them off so the plant produces more leaves. Feature fennel leaves with pork, veal, and fish, and also in sauces and stuffings, as well as an addition to salad dressings.

HORSERADISH: *COCHLEARIA ARMORACIA*

Horseradish is a perennial plant native to western Asia and southeastern Europe, and it has been long known in the Holy Land. Other botanical authorities consider it indigenous to the eastern parts of Europe, ranging from the Caspian Sea area through Russia and Poland to Finland.

The long, white, cylindrical roots produce a 2- to 3-foot high stem in the second year with large basal leaves. In the United States alone an estimated 24 million pounds of horseradish roots are ground and processed annually to produce approximately 6 million gallons of prepared horseradish. Veteran gardeners know, however, that supermarket brands of the herb, whether they are cream-style, horseradish sauce, beet horseradish, or dehydrated horseradish, can't compare with growing and grinding the fresh picked plant from the garden.

Other names for horseradish include mountain radish and red cole and the French call it *Moutarde des Allemands*. Horseradish has been in cultivation from the earliest times but its actual origin is obscure. One authority theorizes that it is possibly a cultivated form of *Cochlearia macrocarpa,* a native of Hungary. The present botanical name, *Cochlearia armoracia,* was given to horseradish by Linnaeus. Linguists explain that the Latin word *cochleare* was the name of an old-fashioned spoon that the long horseradish leaves resemble. It should be noted that horseradish root was included in the *Materia Medica* of the London pharmacopoeias of the eighteenth century under the name of *R. rusticanus.*

Pliny knew it under the name of *Amoracia.* Other plant experts suppose that horseradish may actually be the wild radish, *Raphanos agrios,* mentioned in Greek texts. Others suggest that horseradish may have been one of the five bitter herbs that the Jews were instructed to eat during the Feast of Passover. Having reviewed many different translations of the Scriptures and various versions, I cannot agree that it was meant as a "bitter herb."

Attesting to America's love for horseradish is the annual May International Horseradish Festival in Collinsville, Illinois. Events include a root toss, a horseradish-eating contest, and a horseradish recipe contest. Begun in 1988, the festival was designed to create national awareness for the herb and to the area, where 60 percent of the world's supply is grown. Collinsville and the surrounding area are part of what is known as the "American bottoms," a Mississippi river basin area adjacent to St. Louis. There the soil is rich in potash, a nutrient on which horseradish thrives. German immigrants to the area began growing horseradish in the late 1800s and passed their growing methods from generation to generation. The area's cold winters provide the required root dormancy and the long summers provide excellent growing conditions.

Horseradish is actually a member of the mustard family, which it shares with its gentler cousins kale, cauliflower, and the common radish. Curiously, the bite and aroma of the thick, fleshy white horseradish root are almost absent until it is grated or ground. During this process, as the root cells are crushed, volatile oils known as "isothiocyanate" are released. Vinegar stops this reaction and stabilizes the flavor. For milder horseradish, add vinegar immediately.

Both the root and leaves of horseradish were universally used as a medicine during the Middle Ages. It also was used as a condiment in Denmark and Germany and it was known in England as "red cole" circa 1550. By 1640, Britisher John Parkinson described the use of horseradish as a sauce "with country people and strong labouring men in some countries."

Today, the popular name simply means a coarse radish in contrast to mild edible radish that we grow in our vegetable gardens for salads.

Plant Profile

This herbaceous perennial is often grown as an annual to obtain the best quality root and flavor. Older roots tend to be hotter but coarser and less useful for making sauces. The long and white tapering root of horseradish produces a 2- to 3-foot tall stem in the second year. Leaves are abundant. The lower ones are long stalked and oblong and can be 1 foot in length. Upper leaves are smaller and lanceolate. Flowers are small and white in terminal racemes. Fruit, or seeds, are ovoid to elliptical capsules. Flowers form in midsummer but the root is really the only part of this herb that is used. *Armoracia rusticana* is a useful special variety.

Growing Tips Horseradish roots grow fine in the garden if a plot of tilled ground is enriched with composted manure 12 to 18 inches deep. Horseradish plants require very rich, high organic soil or they will not thrive. Preparing soil well the year before and then retilling with more compost provides the rich growing ground these plants need.

Most herb experts recommend planting with root cuttings obtained from mature plants. These should be straight young roots about 7 to 9 inches long with a bud or growing point at top. Because roots grow both out and down, space them 15 to 24 inches apart. Because they can be invasive, pick a spot where this herb won't intrude on other valued herbs or garden plants. For finest horseradish, remove weeds, mulch with composted manure, and keep well watered.

Harvest roots within a year or leave them in the ground like parsnips. Some gardeners believe this improves their flavor. Roots also store well in a crisper drawer of the refrigerator.

LAVENDER: *LAVANDULA AUGUSTIFOLIA*

Just mentioning "lavender" brings to mind images of perfume and romance. Early written records indicate both the Greeks and the Romans scented their soaps and bath water with this aromatic herb—the name comes from a Latin verb that means "to wash."

In the Middle Ages lavender was thought to be an aphrodisiac, which would attract the opposite sex. Others, according to ancient written folklore, saw lavender differently. A sprinkle on the head of a loved one, a legend tells, would keep the perfume wearer chaste.

Herbalists viewed lavender as having the ability to soothe troubled minds and calm worries. For some it was an alternate to smelling salts and often an ingredient in them. Over the centuries, lavender's distinctive fragrance has continued to be prized in sachets for scenting rooms and clothing, and even comforting the ill in sickrooms and in the homes of the elderly.

For many Americans lavender conjures up the image of old English gardens. For others the picture is of the so-called lavender Alps of southern France, where hills are alive with the wild herb. To most people lavender brings to mind perfume, and it indeed has been a mainstay of the cosmetics industry for centuries. Although there are some records of medicinal uses, the only clear modern one was the use of lavender as a wound disinfectant up to about the time of World War I.

Earlier, British herbalist Maude Grieve noted its use with regard to indigestion. More commonly, in Victorian days lavender was an ingredient used with spirits of ammonia to relieve fainting spells.

Lavender actually is a shrubby plant native to the mountainous regions bordering the western half of the Mediterranean. It is widely cultivated in France, Italy, England, and Australia. The fine aromatic smell is found in all parts of the shrub, but the essential oil is produced from the flowers and flower stalks. Lavender is often sold as bunched, and dried lavender and flowers are used powdered in sachets and for making potpourri.

For the preparation of commercial oil, *Lavandula vera* is used in greatest proportion, despite other lavenders being included as well. The *Lavandula vera,* or English lavender, grows abundantly in sunny areas in many Mediterranean countries. The British, however, are adamant that it grows best and to peak aromatic perfection only in England, although the plant is not authentically English and didn't arrive in the country until the 1500s.

Wherever it is grown, *Lavandula vera* is narrow leaved and grows 1 to 3 feet high. It has short, irregular, much-branched stems that are covered with a yellowish gray bark. The plants have numerous, erect, slender, broom-like branches covered with fine hairs. Leaves are opposite and linear and when full grown are 1½ inches long and green. Flowers are bluish violet and produce the young shoots on long stems.

French lavender oil is made from two distinct plants that are found in the mountain districts of southern France. These are included under the name of *L. officinalis* as well as *L. vera*. To further complicate the picture, Spike lavender, *L. spica,* is also grown. It is a coarser, broad-leaved variety of lavender also found in the mountain districts of France and Spain. According to perfume specialists, the flowers of Spike lavender yield three times as much of the essential oil, known as Spike oil, as can be extracted from the British narrow-leaved plant, but it is considered of lower quality. It should be noted that it has been called by others *Nardus Italica,* which has led several Biblical botanists to believe that this is the "Spikenard" mentioned in the Bible. That has not been proved to my satisfaction.

Further, there is another species of lavender, *L. stoechas,* known also as "French lavender." It is an attractive small shrub with narrow leaves and dark violet flowers. Because it is a wild inhabitant of coastal areas and abundant on some islands, this may have been the original lavender used by the Romans as a perfume for their baths. This species is plentiful and free in Spain and Portugal and was used to scent churches during special days.

Herbal purists say the aroma is more like rosemary than more classic lavanders. America has also come up with its own variety, Lavender Lady.

Plant Profile

Generally, lavender is a bushy and branching shrub. The stems of mature plants often become a dense and woody intergrown tangle. Mail-order catalogs offer descriptions of different types. Leaves typically are opposite, smooth edged, and somewhat hairy. They grow 2 inches or longer, depending on the species, and the plants usually grow to 3 feet tall. Blooming time is June and July. Flowers are small, lavender purple, with 5-lobed corolla and 5-toothed calyx with whorls of 6 to 10 flowers on terminal spikes that are on 6- to 8-inch-long stalks.

Growing Tips Lavender is easy to grow in good, light sandy loam, well-drained, and fairly rich garden soil. It prefers full sun but tolerates light shade. Best bet is to plant cuttings or started plants. Place cuttings in pots with soil mix and keep watered until they set their new roots, or buy started plants to ensure the desired type. Or, order the all-America variety, Lavender Lady, from a mail-order catalog and follow the planting directions. There are options to exercise on species but for growing success, the all-America variety offers assurance and does represent the Biblical herb in a garden.

When lavender plants mature, pick leaves and flower stalks to make decorative scented wreaths, sachets, or potpourris. Hanging bunches in a warm, dry, shaded place to let flowers dry well ensures longlasting lavender. Some prefer using lavender as air fresheners for their homes.

LOVAGE: *LEVISTICUM OFFICINALE*

Whether it is known as Old English lovage, Italian lovage, Cornish lovage, or just good old lovage, the herb is a native of the Mediterranean region and the Holy Land. It still can be found growing wild in the mountainous districts of the south of France, in northern Greece, and in the Balkans.

Lovage (the Latin name, *Ligusticum,* is said to come from the area where this herb abounds, in Liguria) is a true perennial and is relatively easy to maintain as part of an herb or perennial garden. Propagate it easily by root division, the same as with rhubarb, and it will prove a hardy plant. As a stout umbelliferous plant it is described by some herbalists as looking like angelica with thick carrot-like roots. It has dark green leaves on erect stalks that somewhat resemble celery. This herb has an aromatic smell, especially when the leaves are crushed or bruised. The flowers more closely resemble fennel. The distinctive airy leaf and attractive flowers do qualify it as a garden ornamental.

Although ancient herbal books, and some modern reference texts, indicate that lovage enjoyed popularity during the Middle Ages, there are few records that document its uses. One record relates that it had supposedly been a folk cure for a variety of ailments from headaches to colic. One theory is that its medicinal reputation may be traced to its powerful and pleasant fragrance, although it has fallen out of favor today. Noted herbalist Culpeper, nevertheless, has left notes that claim lovage was helpful for digestion. Other writings say that the emperor Charlemagne included lovage in his herb gardens.

Plant Profile

Lovage is a perennial herb with hollow stems that are ribbed like celery stems and divide into branches near the top. The plant can grow up to 5 feet tall. Leaves are dark green, opposite, and compound and glossy. They decrease in size toward the top. Leaflets may be toothed and wedge shaped. Flowers are tiny and yellow in compound small umbels that are 2 to 4 inches across. The seeds are about ¼ inch long, grooved, and aromatic as the foliage is when crushed. Blooming time is usually June and July and the plant thrives throughout North America as a hardy perennial.

Growing Tips Lovage is easier to grow in a moist, fertile, and well-drained soil in full sun to partial shade. Once established it requires little attention. This herb dies back naturally each winter but comes back bigger the following year from its sturdy roots. Once planted, lovage is like rhubarb and it will persist for many years, so pick the right spot for this permanent part of an herb garden or landscape.

A plant or two will serve as a representative of a Holy Land plant. Set a root division with 2 to 3 feet growing room around it. To encourage bushy growth, snip off flowers and trim young stalks periodically. Adding compost mulch around the plant encourages tastier growth but it is best to hold off for a year or so before harvesting or trying this herb in any recipes. Cut leaves and stems to add to soups and stews, and many herbalists recommend drying it by hanging it upside down in a warm, shady spot. Then store dried leaves and stems in tightly sealed opaque containers. Light causes yellowing. Seeds can be gathered when these fruits begin to ripen in late summer.

MARRUBIUM (HOREHOUND): *MARRUBIUM VULGARE*

My memory vividly tells me that this is the herb that gives us horehound candy—of course, it is also known as "marrubium." Actually the herb has many common names dating back centuries including common horehound, hashishat, houndsbane, white horehound, and wild horehound. The plant name for all of them is *Marrubium vulgare,* which has its origin in the Holy Land and today ranges through Europe to Kurdistan, from Mexico to North America.

Horehound is a powerful annual herb with amazingly deep roots. It survives in waste places and roadsides and is easy to grow even in dry, poor soil. The Romans celebrated horehound for its medicinal properties. Its Latin name of *Marrubium* is said to have come from Maria urbs, an ancient town of Italy. That's one theory.

Other authors allege that its name is from the Hebrew *marrob* which translates to "a bitter juice," and some debate whether it was one of the bitter herbs that Jews ate on the Feast of Passover.

Egyptian priests called this plant the "seed of Horus," or the "bull's blood." Horus, from which the plant takes its name, was the Egyptian god of sky and light. Supposedly horehound was a principal ingredient in Caesar's antidote for vegetable poisons. According to old British tales, the herb was brewed and made into horehound ale and taken as an "appetizing and healthful beverage" by workers in parts of England, where it was a popular culinary herb.

In ancient Greece it was credited with curing the bites of mad dogs. Other legends say horehound was used to relieve hepatitis, typhoid, worms, and bronchitis, among other ills. Still others claimed horehound would ward off evil spirits. Horehound has been used through the years in syrups and cough drops, as early as the 1600s in England, and by the Shakers in America for sore throats. Today, some herbalists use it for horehound tea—and horehound candy can still be found.

Plant Profile

White horehound, the type to grow, is an herbaceous perennial. Horehound plants are bushy and have wooly stems and leaves. A member of the mint family, plants produce numerous branching and typically square stems. These are usually a foot or more in height; the plant grows 2 to 3 feet tall. Leaves are opposite in pairs. Lower ones are stalked while uppers are stalkless. They are 1 to 2 inches long, wooly, and round to oval with serrated edges. White flowers appear in dense whorls and have tubular corolla. Seeds form inside small nutlets at the base of the calyx. These are barbed so they will stick to clothing and animal fur to spread the plant around. Horehound blooms from June to September.

Growing Tips Horehound prefers full sun and deep, well-drained sandy soil. White horehound is an easily grown hardy plant—from seeds sown in spring, from cuttings, or by dividing the roots. After seedlings have rooted, transplant them 8 to 10 inches apart. Dividing roots is the easiest method to start new plants from parent stock. Once planted, no further work is needed except weeding. Note that horehound doesn't bloom until it is 2 years old.

MINTS: SPEARMINT, *MENTHA VIRIDIS;* PEPPERMINT, *MENTHA PIPERITA*

Mints have been a treat to gardeners since time began. They have retained their popularity, with spearmint and peppermint being the most popular. Other variations are available so begin with one of these and expand your growing horizons in future years. All have their special tastes, history, and traditions. Spearmint is also called garden mint, Our Lady's mint, sage of Bethlehem, lamb mint, green mint, and more exotically, Menthe de Notre Dame. It's a native of the Mediterranean region and was introduced into Europe and Britain by the Romans. Mint was widely acclaimed by the Romans and the early naturalist Pliny waxed eloquent about it: "The smell of Mint does stir up the minde and the taste to a greedy desire of meate."

Mint history is rather dramatic in early writings. In Greek mythology, one story tells of two strangers who were walking through the countryside and were ignored by most people. As the legend goes, an elderly couple took pity, cleaned their dining table with mint leaves, and cooked them a meal. To their surprise the strangers were the gods Hermes and Zeus who rewarded the couple richly by turning their home into a temple. From that day and legend mint has been associated with hospitality. In other records we find that the Greeks used mint in their temples and rites and also in their food preparation.

Many other references to mint avail in old writings. One of the most significant in the Biblical context is the payment by the Pharisees of tithes of mint, anise, and cumin, which verifies that this herb has been greatly praised and esteemed for many centuries. The Ancients also used mint to scent their bath water. Tracing mint migration finds that it was cultivated in the monastery and convent gardens of the ninth and tenth centuries.

Because of its strength, mint seemed to fascinate rulers and poets alike. Even Chaucer refers to "a little path of mintes full and fenill greene" in his writing. Dozens more quotations refer to mint's historic roots. Naturalist Gerard noted its medicinal properties: "They lay it on the stinging of wasps and bees with good success." Nicholas Culpeper notes dozens of maladies for which mint was used.

Consider this amazing fact. In the fourteenth century mint was used for whitening teeth. Today, its distilled oil is still used to flavor toothpastes and is widely used to flavor candies and many chewing gums. Mint is also a popular perfuming agent for soaps. Mint is probably one of the most widely known herbs and is grown worldwide.

Plant Profile

Mints are pervasive, spreading perennials that can take over a garden and crowd out other desired plants. They send up new plants from spreading underground roots. Mint plants are easily identified by their square stems and distinctive smell. There are many types beyond the spearmint and peppermint. However, it is best to keep different species well apart or in different parts of a garden or landscape because they tend to crossbreed and lose their individual attributes.

Plants grow from creeping roots with the erect, square stems maturing to 2 to 2 ½ feet tall. The mint's bright green leaves are opposite, wrinkled, lance shaped, and toothed. Flowers are relatively tiny and are densely arranged in whorls at the axils of the upper leaves. These blooms form tapering spikes that range from pinkish lilac to purplish. Fruits or seeds form 4 smooth nutlets. Flowers form in July and August.

Growing Tips Mints like rich soil in moist areas in full to partial sun. Consider several plantings because mint tends to be used more than some of the other herbs and a steady supply is needed, especially for mint teas. Best way to plant mint is from root divisions. In fact, peppermint is a sterile hybrid that doesn't produce seed, so cuttings or divisions are the only means by which to grow the herb.

Make root divisions in early spring by dividing parent plants, or purchase specific mints. Place firmly in the soil in a moist area and cover with 2 inches of soil. Set plants 6 to 8 inches apart. Cuttings can also be made in summer, just root them and plant in a garden. With pervasive mint, some herbalists advise planting in large pots or tubs sunken in the ground so that they won't spread. I've used 5-gallon buckets with bottoms cut out and that keeps mint in one area effectively.

Cut mint whenever the need arises. It is best to cut sprigs a few inches above the root, on a dry day, at least after the dew has disappeared and before the hot sun has taken any oil from the leaves. Use for seasoning or crushed in iced tea, for example. To save mint leaves for winter, tie sprigs loosely into bunches and hang to dry like for other herbs.

My wife and I love making mint sauce. We strip the most tender leaves from stems and chop in a bowl. Then we add vinegar, sugar, and water and pour into glass bottles that can be securely sealed. That gives us our supply for the tastiest lamb dinners ever.

More Mint Facts

Other tasty mints include apple mint, curly mint, pennyroyal, and peppermint. Apple mint, *M. suaveolens,* is a hardy perennial and a very robust grower with interesting light green leaves that are somewhat hairy on the upper surface and downy underneath with serrated edges. They can grow up to 3 feet tall.

Curly mint, *M. spicata* (variety crispii), is a plant valued more for its intricately curled, fringed leaves than for its fragrance. The foliage is dark green and splotched with a contrasting lighter green. They can grow up to 2 feet high and are invasive.

Pennyroyal, *M. pulegium,* is a creeping mint that can be a ground cover in cool climates. It has many stems, which grow up to 12 inches tall, covered with small, round to oval, dark green leaves. Pennyroyal has a lemony aroma and is used to flavor meat, puddings, and fish entrees.

Peppermint, *M. piperit,* is a spreading plant with a penetrat-

Aloe

Borage

Caraway

Chamomile

Chives

Coriander

Cucumber

Dandelion

Dill

Endive

Fennel

Garlic

Horehound

Hyssop (Roman)

Lavender

Marjoram (wild)

Melon

Onion

Oregano

Parsley

Rosemary

Saffron crocus

Sage (pungent)

Savory (summer)

Thyme

ing yet pleasant, mint aroma. It grows 1 to 2 feet high, but can reach 3 feet when in bloom. Lance-shaped leaves are deeply notched when mature and flowers are purplish.

MUSTARD: BRASSICA NIGRA

> So Jesus said to them, "Because of your unbelief; for assuredly, I say to you, if you have faith as a mustard seed, you will say to this mountain, 'Move from here to there,' and it will move; and nothing will be impossible for you."
> —Matthew 17:20

The mustard being discussed in the passage is probably Sinapis. Botanists say it closely resembles black mustard. Other authorities, however, believe the plant in question is the Khardal of the Arabs, which is a tree found near the Sea of Galilee. That tree bears numerous branches and has small seeds with a similar flavor to mustard.

Black mustard is also said to have been eaten by the Romans as a green vegetable. They also pounded the seeds into a flour and mixed it with wine to make what may have been the earliest known table mustard. Mustard, however, was not only esteemed for its spicy flavor, but also for medicinal uses. Early herbalists in Europe also recommended taking mustard to treat toothaches. Records from America's colonial days state that settlers and Indians used cultivated mustards for food and medicine as well. Black mustard seeds were ground into powder, mixed with animal fat, and used to relieve pains of sprains and strains. Another report says some Indians used the mustard to relieve toothaches, too.

Mustard gets its name from *mustum*, which translates to "the must," and *ardens*, which signifies "burning." It was originally eaten whole or slightly crushed according to old writings. Shakespeare mentions mustard more than once and specifically names Tewkesbury mustard in *Henry IV*.

Over the centuries mustard was used in poultices and in

other medicinal ways. Many seasoned senior gardeners can probably recall the mustard plasters used by parents to cure colds and congestion. Those time-honored remedies made the skin feel warm and opened the lungs for easier breathing. No doubt many thousands were treated to the old-fashioned mustard plasters until they were replaced by the invention of heating pads and menthol vapor medications.

The first honey-mustard vinegrette was made with honey and vinegar and a little cinnamon so it would keep until it was mixed with more vinegar for use. Mrs. Clements, of Durham, England, sold mustard in balls for such recipes at the close of the eighteenth century and seemingly invented the method of preparing mustard flour, which was sold under the name of Durham Mustard for many years. At Dijon, where the best continental mustard is made, the condiment is seasoned with various spices such as anchovies, capers, tarragon, and walnuts or mushrooms.

Today, the black mustard grows throughout Europe to Asia Minor, in Northern Africa, and is naturalized in North and South America.

Plant Profile

Black mustard is an erect annual maturing to 3 to 6 feet in height with smaller flowers than the white mustard. Leaves are alternate and various shapes. The lower ones are pinnately lobed and coarsely toothed. Upper ones are less lobed. Flowers are small and yellow with 4 petals that form a cross. The distinctive, rounded Maltese cross is unique. Typically about 10 to 12 dark brown or black seeds form in the smooth, flattened pods after flowering time, which for this annual is early summer. Black mustard seeds are about half the size of white mustard seeds.

Growing Tips Mustard is rather hardy and grows well from seed sown in spring. Plant tiny seeds ⅛ inch deep in a sunny area in rich, well-drained soil. Thin to 8 to 10 inches apart, keep weeded, and fertilize regularly during the season because mustard uses up nutrients more quickly than other herbs. Although an annual, mustard will self-seed and can overgrow areas. Just pull out extra plants as necessary.

Harvest seeds when pods have turned from green to brown before they split open and spill the seeds. Pull plants, spread on a sheet, and let seeds fully ripen so pods shatter open easily. Keep mustard seeds in tightly closed jars.

OREGANO: *ORIGANUM VULGARE*

Oregano most likely conjures images of tasty Italian food, tomato sauce, and pizza. That's natural. Oregano is one of the more popular flavorings that comes from the Mediterranean Sea area. Oregano's name actually translates from the Greek to "joy of the mountain." Early in its history, oregano was praised by Roman scholar and writer Pliny and Greek physician and writer Dioscorides. In those times, oregano was recommended for medicinal purposes in that part of the world. According to ancient texts, the Greeks made poultices from the leaves to cure sores and to help aching muscles. In the sixteenth century, John Gerard included the plant in his herbals. Colonists brought oregano to North America where it escaped from gardens and found the climate to its liking; it now grows wild in many areas.

Doing more extensive research to trace it as a plant of the Holy Land, I found much confusion and debate about the origin and history of this herb or group of related herbs—and I don't count it as a true Biblical herb. Nevertheless, consider it as a basic herb for herb gardens and in culinary projects if you wish.

PARSLEY: *PETROSELINUM CRISPUM*

Botanical scholars believe parsley probably originated in Lebanon while others note that parsley grows native from Sardinia to Lebanon and now is a popular herb in temperate climate areas of the world. Most of us think of parsley as an attractive green garnish, used often by restaurants to dress up a plate of food, and simply discard it. Perhaps this chapter will reveal some of its values and encourage you to grow and use it more widely.

Parsley is a marvelous breath freshener. That was its reputation in ancient Roman days when old texts reveal that parsley was used to conceal the smell of alcohol. From that ancient lore, kindly bartenders today often pass along the wisdom of the ages, "Try a bit of parsley before you go home and she'll never know you've had a drink." It isn't quite that effective but no doubt has been a noted breath freshener for centuries.

Greek customs seem to have associated parsley with death and parsley was used in ancient burial rites. This may be related to parsley's ability to deodorize. One legend tells us that parsley sprang up where the blood of the Greek hero Archemorus had fallen. Still other writings show that parsley was used in wreaths and as part of awards for outstanding athletic ability. History lessons reveal that Hercules supposedly had picked parsley to garland his head.

Several types of parsley are available: from the common, plain-leaved to the curly or broad-leaved. Best bet for its early significance is the variety crispum, which was even mentioned by Pliny. By the Middle Ages parsley had become part of herbal medicines. Folklore relays tales of parsley being used to cure such ailments as the plague and jaundice. Some facts from modern research are worth reviewing as well. Parsley is a major vitamin source. Parsley contains vitamin A, more vitamin C per volume than a typical orange, several B vitamins, plus calcium and iron. Actually, parsley is more widely used in European meals than in America. The French feature parsley with ham, use it with garlic and butter, as well as with grilled meats and poultry. In Spain as well as in Mexico parsley is a key ingredient in salsa verde.

Plant Profile

Two types of common parsley are available for the garden: the flat leaf and curly leaf. Distinctive leaves are divided pinnately into feathery sections and the curly type as curls. The flowers are tiny and greenish. Small seeds are oval, ribbed, and grayish brown. Plants mature about 12 to 18 inches if given sufficient room. Parsley can be an attractive border or edging plant for its interesting foliage combined with other flowering herbs or compact flowers as mix-and-match gardens.

Growing Tips Parsley thrives in moderately rich, moist, and well-drained garden soil and a partially shaded position is best. Be especially patient with the seeds. Soak them overnight before planting when garden soil is well warmed in spring. Be sure to keep the planting area moist because it may take several weeks for seeds to sprout. Try transplanting young, year-old plants too. If plans call for parsley often, plant a row or two and some in pots or containers for indoor growing and use. When seedlings are well up, thin to 8 inches apart, or 2 seedlings in a pot. For most succulent parsley be sure to water regularly during dry periods and also keep well weeded for best growth. Some herbalists prefer several sowings of parsley. By the second year parsley sets its seeds and will self-sow for more plants the following year. Remove all flower stalks and prune parsley periodically to encourage good growth. To harvest for salads, cooking, or garnish, simply snip parsley with scissors. Dry parsley in a shady area, crush it, and store in airtight containers, or grow fresh indoors in a pot for a plentiful supply of this tasty herb all winter.

POT MARIGOLD: CALENDULA, *CALENDULA OFFICINALIS*

Some botanists prefer that this herb be properly named Calendula and in that order in a book. Perhaps so. However, so many people call this pot marigold that it seemed appropriate to include it here under that popular name. Actually it has other names from garden marigold to holigod and Mary's gold. This plant, by whichever name preferred, does indeed trace back to the Holy Land and many of the Mediterranean Sea countries.

Many herbal purists prefer to think of this as a flower, but essentially it is a flowering herb that at one time was much more popular than it is today. In olden times calendula flowers were used to reduce inflammation and to heal wounds as well as act as an antiseptic. Clinical studies have suggested that marigold flower extracts lower blood pressure and have sedative effects. A news item in 1995 reported that an Australian patent had been issued for the use of marigold extracts in the treatment of burns in humans.

The Romans actually gave the plant its name. Legend has it that because the plant tends to bloom in the first part of every month Roman logic decided that "calends" would be an appropriate name, hence *calendula*. It does bloom again and again and again to keep all gardeners happy with its beauty.

Ancient herbalists felt that because of its color caldendulas might be useful to treat jaundice. Other folklore promised that using "marygolde" would give a person special visions, including elves and fairies. One folktale says that using calendulas with other herbs in a secret mixture would enable a young woman to see her true love. It is assumed that when Shakespeare mentioned "Marygold" he actually meant *Calendula officinalis* because of its wealth of bloom. Old English authors called it "golds" or "ruddes," and later the plant was associated with the Virgin Mary.

Despite many bits of lore linked with magic, calendulas never achieved much notoriety as a medicinal herb. Some records from America's Civil War indicate that calendulas were used to promote healing of wounds, but in those horrible times just about every idea was being tried. On balance, calendulas were much more widely used for their flavor and cooking. Old cookbooks and texts tell how calendulas were used almost like vegetables. John Gerard, writing in the sixteenth century, indicates that calendula petals were a common ingredient in soups, puddings, and even wines.

Plant Profile

Calendula is an erect, many branched annual plant covered with fine hairs, which grow 1 to 2 feet tall. Leaves are alternate and oblong with smooth to faintly toothed edges. Ray flowers appear regularly in solitary terminal heads 1 ½ to 4 inches across. They are pale yellow to deep orange and close at night. Plants bloom profusely from June to October. Seeds are boat shaped and light yellow. Although a native to southern Europe, the Middle East, and Africa, calendulas flourish in cool, temperate climates. Petals have a pungent flavor. Today, plant breeders are working to produce more dramatic blooms, vivid colors, and bloom abundance. In the past two decades, these flowering herbs have won many new friends. Besides more profuse blooms on new varieties, today some are fully double and range in color from creamy to bright orange. They dry well, too, for colorful herb and flower arrangements and wreath uses. Check mail-order catalogs to find new varieties worth a try.

Growing Tips Calendulas are easily grown from seeds. Sow seeds directly in the herb or flower garden ground in April or May when soil is warm. Weed well and keep moist until seeds sprout. Thin to 8 to 10 inches apart when seedlings are well rooted. These herbs prefer average to rich soil in full sun. Adding organic matter seems to help them prosper and bloom more profusely. Add compost as mulch to stop weeds and improve the soil. Water well during day periods to keep them flourishing. These plants are hardy annuals and will hold up well into cooler weather. If treated well, they will seed themselves to keep producing more plants year after year.

To harvest, just pinch the flower head off the stem and pull each petal and dry them on paper or cloth in a shady area. The petals, with their slight aromatic bitterness, are used by herbalists in fish and meat soups as well as in rice dishes. Use the whole flowers as garnish as people did in medieval times.

ROSEMARY: *ROSMARINUS OFFICINALIS*

A symbol of fidelity and remembrance, rosemary was used in the holiest of Christian ceremonies, weddings, and funerals. For many centuries people thought that the rosemary plant would never grow taller than 6 feet in 33 years so that it would never stand taller than Christ. Another wonderful story is that Mary once spread her cloak to dry on a rosemary bush while fleeing from Herod's soldiers with the Christ Child. After that the flowers were forever blue instead of white.

Another legend relates that when a rosemary bush grows vigorously in a family's garden it means that a woman heads the household. For many decades rosemary was one of the early bathing balms. People would draw a full tub, relax in it, and soak with the piney fragrance of rosemary around them.

A deeper investigation into herb history reveals that rosemary was alleged to have been useful to protect against evil spirits. In the Middle Ages people placed sprigs of rosemary under their pillows for protection and to prevent bad dreams. Queen Elizabeth of Hungary bathed in rosemary water and remained beautiful into her 70s, and her formula, in her own handwriting that is dated 1235, is preserved in Vienna. During the sixteenth century, rosemary was used as an air freshener by mixing rosemary leaves with sugar, heating the concoction, and letting it filter through the home.

Herbalists also prescribed the herb for the treatment of depression, headaches, muscle problems, and a variety of other ailments.

St. Thomas More has been quoted as saying, "I lett it runne all over my garde wall, not onlie because my bees love it, but because 'tis the herb sacred to remembrance, and therefore to friendship . . ." Rosemary had been favored for use in wreaths worn by brides and also as part of bouquets they carried. It also had a place during funerals. Old records say that rosemary sprigs were often thrown into the grave during burials as a symbol of friendship for the departed.

Today, rosemary is cultivated in Europe and North and South America. In Latin the plant was called *rosmarinus*. Most sources interpret it as *ros marinus* which translates "dew of the sea." According to one unique herbal record the fragrance is so strong that it can be smelled miles out at sea when it is being harvested on shore. In its native habitat rosemary actually grows at low altitude and therefore near the sea. Rosemary does not lose its flavor by long cooking as other herbs do. Fresh leaves have a purer fragrance and are best.

Plant Profile

Rosemary is an evergreen perennial shrub with ash-colored scaly bark and green needle-like leaves. It has a grayish green appearance and matures 5 to 6 feet tall, so it needs lots of room to grow properly. At Meadowsweet Herb Farm in Vermont, I recently saw a rosemary plant that was 5 feet tall and almost 6 feet across. In restricted container culture it will grow 2 to 3 feet tall. Leaves look like needles and are $1/3$ to $1\,1/2$ inches long and grow opposite. They are narrow and leathery with the upper portion dark green; underneath they are white and somewhat hairy with a piney fragrance. Abundant flowers are $1/2$ inch long and colored lavender to pale blue. Flowers grow in clusters of 2 or 3 along branches. Fruit (the seeds) are very small, spherical, smooth nutlets. Rosemary tends to bloom in February and through the spring. Newer cultivars bloom abundantly and can be found in varieties in mail-order catalogs. Dwarf types such as Prostratus, Collingwood Ingram, and Majorca Pink grow only 2 feet tall.

Growing Tips Rosemary can be a bold accent plant in the garden as a distinctive, aromatic shrub or one of the low-growing cultivars. Rosemary does best in a light, rather dry soil in a sheltered situation spot, out of the wind with southern exposure for adequate sun. Seeds may be planted in a sunny spot but rosemary is best propagated by cuttings, layering, or root division. Take 6- to 8-inch cuttings in late summer and plant in a shady border, two-thirds of their length in the ground. By next summer they'll be well rooted and ready to transplant to their sunny permanent home. Grow new plants by layering. Simply insert a few lower branches into soil and hold with a rock until roots form and then cut away from the parent plant and transplant to a desired location. Older plants are deep rooted and difficult to transplant. Some herbalists prefer root division to get new plants. Rosemary thrives under drier conditions than many herbs and enjoys a bit of lime and organic matter added and scratched into the soil around it. In future years add some fertilizer to keep it growing well, according to directions on the brand used.

Pick rosemary sprigs all year long. Cut 4-inch pieces from tender tips of branches for use. Freeze extra rosemary but note that it becomes stronger when frozen. Cooks use rosemary for fish, meat, and poultry as well as mixed into vegetables. It is often recommended for potatoes, zucchini, and tomatoes. Try it with other vegetables and in marinades and salad dressings, too. This is one of the herbs that seems to be gaining popularity faster as more people discover its merits. Some herbalists say that sprinkling rosemary leaves on glowing charcoals during grilling gives a pleasant effect. Be aware that rosemary is among the most powerful herbs, so small amounts are sufficient as taste testing will affirm.

RUE: *RUTA GRAVEOLENS*

Rue has long been a symbol of repentance and sorrow. It is also called "herb of grace" because in early Christian times it symbolized the grace given by God following the repentance for one's sins. History affirms that brushes made from rue were once used to sprinkle holy water at the ceremony preceding High Mass. In Luke 11:42 we read: "But woe to you Pharisees for you tithe mint and rue and every herb, and neglect justice and the love of God; these that you ought to have done, without neglecting the others." Rue was widely used as food flavoring and as medicine during the New Testament period and found favor as one of the herbs collected as part of taxing.

Rue can be found in many different historical and herbal references. The Greeks and Romans believed it protected them from a variety of diseases; in fact, there are records that they spread it on floors of public buildings and also walked outside carrying bouquets of rue to protect them from disease. Today this hardy evergreen shrub grows in ditches, on hillsides, and in waste areas in Mediterranean countries and the Holy Land, just as it grew in Biblical times.

Rue has a long, colorful history and some of the more unusual legends attached to it. The name "ruta" comes from the Greek word *reuo*, which means to "set free." Linguists say that probably relates to the herbs supposed value to help cure diseases. One researcher noted that Aristotle had said rue eased nervous indigestion. In the process, rue also came to be known as an anti-magic herb to stave off witchcraft, a belief that continued through the Middle Ages.

Pliny thought rue to be of use for the eyes and indicated that painters and artists used it to sharpen their sight. Other early records cite rue as an antidote to poisons and bites by insects.

Turner's *Herbal,* circa 1560, also indicates that rue is one of the best known and most widely grown herbs for medicinal uses and to aid sight. Herbalist Nicholas Culpeper recommended it for sciatica and pains in the joints. Even William Shakespeare refers to rue in Richard II: "Here in this place I'll set a bank of rue, sour herb of grace; Rue, even for truth, shall shortly here be seen, In the remembrance of a weeping queen."

From the earliest times rue has been regarded as useful in warding off contagion and also in preventing the attacks of fleas and other noxious insects. One old English record reveals that judges had rue on the court bench to prevent possible infection brought by sick prisoners into court. Perhaps just as important was the aspect of preventing attacks of fleas and troublesome insects. Remember that bugs were a problem in ancient times and people realized that they carried diseases with them. Rue-water sprinkled in the house "kills all the fleas," according to one source.

Plant Profile

Although rue isn't one of the more attractive herbs, its distinctive blue-green foliage and yellow blooms do have their place in a Biblical garden. New cultivars such as Variegata and Jackman's Blue with its blue foliage can be appealing. Rue is a perennial evergreen shrub-like plant with wood stems at the base and herbaceous growth higher up. Leaves are alternate, pinnate, and 3 to 5 inches long. They are somewhat fleshy, blue-green, and often covered with a powdery material that has a pungent odor. Flowers are yellow to yellowish green, ½ inch wide with toothed, concave petals in loose clusters at the top of the plant. Blooming season is usually June to August. Rue grows about 3 feet tall. Because of its unique color foliage and shrub growth it can add another dimension to container gardening.

Growing Tips Grow rue from seeds, cuttings, or root division. Start seeds indoors in peat pots and transplant outdoors when soil is well warmed. Rue likes full sun and well-drained and slightly clay loam soil. Space plants 15 to 18 inches apart. Root cuttings in the shade and then transplant to their permanent home outdoors. Once established this shrubby perennial can give years of pleasure. Container-grown rue adds appeal with its distinctive foliage. Pot in container soil mix and place in a sunny window.

Veteran herbalists note that rue may live much longer and be less injured by winter when grown in a poor, dry soil than in fertile ground. So, for gardeners with poor soil areas, give rue a try.

SAFFRON: THE HERB FROM THE CROCUS, *CROCUS SATIVUS*

In tracing the roots of Biblical plants, it is occasionally possible to find an identification on which all authorities are in agreement. Although saffron, in Hebrew, *karkom*, is mentioned only once in the Scriptures along with spikenard and cinnamon, there is no question about what plant is being discussed.

> Spikenard and saffron; calamus and cinnamon, with all trees of frankincense: myrrh and aloes, with all the chief spices . . .
> —*Song of Solomon 4:14*

Saffron comes from one particular crocus, the saffron crocus, *Crocus sativus*. It is native to Asia Minor and Greece and is found in other Mediterranean countries as well. After the stigmas and styles (the reproductive parts of the flower) are gathered, they are dried in the sun, then pounded into small cakes, a process not very different from what was done in Biblical times. They are then used as a yellow dye and also for coloring in curries and oriental foods. Considering that it requires at least 4,000 stigmas and upper portions of the style of the blue-flowered saffron crocus to make an ounce of the yellow dye, it is not unreasonable to understand its expense as a coloring product.

Mentioned also in the Talmud, during Biblical times saffron was scattered during wedding ceremonies and mixed with wine. It has been used to make perfume and to color confections as well. Some texts indicate that saffron was given medicinally as a stimulant and with spices and flower petals as scents to perfume rooms. In other passages of the Scriptures, according to the translation of Dr. James Moffatt, pains are taken to change some of the wording of the Song of Solomon. Where, in 6:4 it is written in the authorized version, "Thou art beautiful, O my love, as Tirzah, comely as Jerusalem . . . ," Dr. Moffatt stated that no love song could compare a young maid to a city. Therefore, he renders the passage as "You are fair as a crocus, my dear, lovely as a lily of the valley." While it may seem more appropriate to compare a young girl to flowers than to cities, other scholars believe that Dr. Moffatt has taken some liberties in his translation.

If, like Dr. Moffatt said, a crocus is a more fitting comparison, it certainly can be justified geographically. Many believe that if the Song of Solomon was actually composed by Solomon, it was done somewhere in the hills and mountains of Lebanon. There, four different species of crocus do occur in profusion. Closely resembling the American springtime crocus, the saffron crocus actually blooms in its native habitat in late autumn. It is a tiny plant with subterranean corm. It produces several narrow leaves and one or more large, bluish lilac flowers. Choose the purple-blue spring crocus as a reminder of the saffron crocus or the purple fall crocus that also thrives across northern regions of America.

Plant Profile

Crocus, a genus which are closely related plants with a single name, are hardy perennial plants and a member of the iris family, Iridaceae. Crocuses produce a single tubular flower from a corm. They have grass-like leaves that grow in a rosette pattern. The common fall crocus, *C. sativus,* has bright lilac-blue flowers. Spring-flowering crocuses include early flowering, so-called Dutch crocus and are available in violet to purple hues and in various other colors. Actually, many types of crocus thrive in the Holy Land as they did before the birth of Christ. White and pink, blue and yellow, as well as lilac and purple crocuses grow in the fields, rocky hillsides, and along roadsides. However, only the blue-flowered ones produce saffron. These must have grown in great abundance to have withstood the constant harvesting, which the demand for saffron creates.

Growing Tips Crocus bulbs are easy to plant. Today's crocus varieties have been specially developed for more profuse blooms and showier colors than the original wild species. Truest in appearance to the Scriptural plants are the blue ones, whether desired for spring or fall bloom. The fall types multiply rapidly without care or trouble. Plant autumn crocuses in the spring, in sun or light shade areas. As a perennial, they become a permanent part of a plantscape beneath trees, under shrubs, in beds, and along pathways. Once planted they increase year after year to produce profusions of blooms.

Plant spring-flowering crocuses between mid-September and early November before the ground freezes. New varieties produced by plant hybridizers and offered by mail-order companies are related to the crocus of the Scripture and offer larger, more abundant blooms. Plant bulbs 3 to 4 inches deep in clusters or groups. Once planted, they are so persistent that even an amateur gardener can obtain perfect results. Most crocuses multiply year after year by forming bulblets from their corms, during the growing season. For best results, obtain larger corms which cost a little more but produce bigger blooms and allow the garden a better start.

SAVORY: SUMMER, *SATUREJA HORTENSIS;* WINTER, *S. MONTANA*

This section features two for one: summer savory and winter savory. These powerful old herbs trace their heritage back more than 2,000 years in texts from both the Old World and New World. The genus Satureia is the Latin name used by Pliny, the famed Roman scholar and naturalist. Although the two savorys mentioned above are the ones to grow, there are about 14 species of highly aromatic hardy herbs or under-shrubs and all, except one species, are natives of the Mediterranean region and Holy Land environs.

Summer or garden savory is an annual and winter savory is a perennial. The annual is more usually grown but the leaves of both are used in cooking, and, like marjoram, are tasty to season dressings for chicken, turkey, veal, or fish.

Historically the genus's Latin name, Satureja, is attributed to Pliny and is derived from the word *satyr* the half-man, half-goat creature of Roman mythology. Romans used savories in their cooking and often flavored vinegars with them. Both species were noticed by the poet Virgil as being among the most fragrant of herbs and he recommended they be grown near beehives. Savory was widely cultivated in remote ages because East Indian spices were unknown then. During Caesar's time it is said that Romans introduced savory to Europe and England where the herbs became popular for cooking and medicinal uses. Reportedly, vinegar flavored with savory and other aromatic herbs was used by the Romans in the same manner as we use mint sauce on lamb and other meats today.

The Saxons actually gave this herb the common name of "savory" for its pungent, dramatic taste. Through the years it became a favorite herb in Italy and is still featured in that cuisine. Elsewhere, gardeners planted hedges and mazes of savory and other herbs during the heyday of the Tudors. Even seventeenth-century herbalist Culpeper wrote about them. He noted savories for their "heating, drying . . . and are good in asthma and other affections." Oddly he also recommended savory as a cure for deafness. Another seventeenth-century herbalist, John Parkinson, wrote about savory with directions for drying and powdering it for use in cooking. Savory, along with mints, marjoram, and lavender, appears in Shakespeare's *The Winter Tale*. American pioneer John Josselyn, one of the early settlers in America, left a list of plants introduced by the English colonists to remind them of the gardens they had left behind. Winter and summer savory are two of those he mentioned in his book, published in 1672. During that time, savory seems to have been trusted as a remedy for indigestion as well as a cough remedy and also as an appetite stimulant. It was also known as a poultice to relieve insect bites and stings.

Plant Profile

Summer savory, *S. hortensis*, is an annual with a branching root system that supports a bushy plant with finely haired stems. All parts of the plant are aromatic. Leaves are soft, linear, and about 1 inch long and hairless. They are attached directly to the stem in pairs and are grayish but turn purplish in fall. This annual grows 1 to 1 ½ feet tall. Flowers are white to pale pink, two-lipped, and ¼ inch long and appear in groups of 3 to 6 from midsummer to frost.

Winter savory, *S. montana*, is a hardy perennial semi-evergreen that is woody at the base and forms a compact bush. It has a stronger fragrance than the summer savory, but the summer type is sweeter and is considered more delicate. This savory has dark green, glossy, and lance-shaped leaves about 1 inch long. Flowers are white to lilac with lower lips spotted with purple. They are ⅓ inch long and grouped in terminal spikes. Winter savory grows 6 to 12 inches tall and blooms July through September.

Growing Tips Both types of savory can be grown from seeds or cuttings. Summer savory seeds sprout rather quickly compared with most herb seeds. Sow seeds ⅛ inch deep in pots or trays to transplant later or directly into an herb garden ground when soil is warm in spring. Space plants 10 inches apart and keep well weeded and watered. Summer savory also responds well to container growing in a pot with proper soil-less soil mix and fertilized periodically.

Winter savory seeds are slower to germinate so start indoors in pots or trays or be patient outdoors. Thin or space plants 10 to 12 inches apart. Both type savories like average to light, well-drained soil and perform best in full sun locations.

Begin harvesting summer savory when plants are 6 to 8 inches tall. Simply snip the tops of branches as needed. When plants flower, cut off whole plants, lay on a clean sheet to dry in a warm, shady spot. Next, strip off dry leaves after a few days and store in airtight glass jars. Harvest winter savory the same way, but leave this perennial in place or pot it up for indoor use in winter in areas with very cold climates. Savory's distinctive taste, somewhat similar to marjoram, does well in stuffings, stews, soups, and with various types of bean dishes. Savory also has a medicinal use: both old authorities and modern gardeners agree that a sprig rubbed on wasp and bee stings gives instant relief.

Chapter Five

Pick the Right Site, Prepare Well, and Plant Wisely

This chapter provides advice on how to select the right location for a Biblical herb garden, prepare and improve its soil, and nourish herbs as well. Well-fed herbs prosper, perform better, and, as with other types of plants, resist insects and diseases. Fortunately, most herbs are hardy and don't have many problems. However, many are slow to start after seeding. Give herbs the best growing conditions and they will offer many rewards.

The first basic step for all gardening success is to pick the right location. Look around your property. Naturally, an herb garden close to the kitchen is especially handy because snipping some herbs to add to food is easier than if the garden is far away on the property. Most herbs like sun, and, generally, need at least 6 hours of it each day. Southern exposures are best for herbs. Next best are eastern, then western. Most also prefer rich, loamy soil. Some can take poorer conditions but it pays to improve soil for herbs and all plants.

Here's how to improve soil. Pick up a handful of rich, warm soil in spring. If it crumbles freely in your palm, you are approaching the ideal texture. Consistency of the growing medium is of underlying importance. The closer you have or can rebuild soil to a granular feel, with clusters of soil that easily shake apart, the better your garden will grow. If your garden does not meet this criteria, it's time to combine organic matter and manure, so that it decays into a valuable soil additive. This process is called "composting."

COMPOST MAKING MADE EASY

Using readily available materials, you can turn organic matter into usable compost in as few as 14 days. Soil, like the plants that grow in it, is alive! Millions of bacteria, fungi, minute animals, and other micro-organisms inhabit it. As good gardeners realize, just as we need food to grow, our plants also need water plus nutrients. Improving soil texture enables plant roots to move more easily through the soil to pick up the nutrients they need to grow well.

Compost piles do not exactly have eye appeal. If you live in an area where homes are close together, then you must consider the appearance of your compost area. Also, decomposing organic matter may not smell nice to you or your neighbors. Select an out-of-the-way location. You probably will want to screen the area from view with a fence, hedge, or shrubs. I use a berry hedge where possible to provide the family with tasty berries as a bonus.

A shady area is good for the composting process, but avoid low areas in which rain collects and the ground remains soggy. While bacteria that help make compost into good humus (well-rotted organic material with a crumbly feel) require ample moisture, they must have air and oxygen as well. That's why turning a compost pile is helpful. It lets bacteria obtain oxygen to do their work better and faster, breaking down organic material.

Next step in preparing compost is construction of the pile. Remember that air circulation is important. The more air that circulates through the pile and around it, the faster the decomposition. It's good exercise but hard work.

Almost any organic material can be composted with the simple Indore composting method. This method was developed and practiced in England by Sir Albert Howard, father of the modern organic movement. It is the best, most widely used method by small and large gardeners and it is both practical and efficient. Basically, the Indore method is the simple layering of various materials.

Begin with a 6-inch layer of green organic material. Don't let the "green" name fool you. Grass clippings or dried leaves may be brown but they are considered "green" matter in the terminology of compost making. Over this first layer of green material—clippings, leaves, old weeds, and vegetation pulled from the garden—add a 2-inch layer of manure. The objective is to add nitrogen to hasten the decay process. Cow, horse, sheep, and poultry manure will do the job. It is best if the manure also includes straw, or shavings. If manure is not readily available obtain dehydrated manures, which are found at most garden centers. These commercially prepared, dried manures are excellent to use in compost making.

The next layer in the compost pile should consist of an inch or two of garden soil, evenly spread to ensure effective interaction. It will have natural and useful soil bacteria in it to help with the composting process. Then add more organic material, weeds, leaves, and kitchen parings. Never use any diseased plants.

A small pile, just 4 feet wide, 4 feet long, and 4 feet high, works well. I prefer to have 2 piles working, one to use and the other "in process."

Keep Compost Piles Moist

As you apply the layers, sprinkle them with water. All organic material going into the pile should be moist, especially when using dry leaves, grass clippings, and other dry materials. Leave a depression on the top so that rain water can be caught and allowed to trickle down through the layers. If you are in a drought period, be sure to water the compost pile once or twice a week.

Two types of bacteria will be at work in your compost. These are called aerobic and anaerobic. The first type, aerobic, needs air circulation in order to perform. The second type, anaerobic, works more slowly and proceeds without much aeration of the pile. If you turn the compost once a week you can expect the process to be finished within two to three months, depending on weather and other variables.

Bacteria are the microscopic helpers in any compost. Soil bacteria cultures may be purchased from various sources including Plants Alive company, a major natural gardening product firm. The easiest way to meet this need is to save the remnants of a previous compost pile. Add the material not quite completely decomposed to the new pile or to the soil or manure layers. Spring and summer are the best times for making compost. Sun and rain, warm weather, and water speed the process.

Field Composting Is Easier

Field composting is easier than the Indore method but takes much longer to produce finished compost. Basically you gather green matter, apply it in layers to a pile or along garden rows, and just let it rot by itself. Keep adding more organic material, raked leaves, grass clippings, weeds, until you have a satisfactory 4- to 6-inch layer. Then add manure or other nitrogen-containing materials from your garden center. I like this system because all I must do is spread organic materials along garden rows and let them rot down in place as a natural mulch that also helps stop weed growth.

Composting is a continual process; the remnants you apply to the newer piles help "inoculate" them with more bacteria to keep the microorganism level in high gear. Just about any type of vegetation and organic matter can be put to use to make compost. You can turn garbage into humus with little effort—vegetable parings, tops of carrots, radishes, beets, corn husks, pea pods, lettuce and cabbage outer leaves—any type of household vegetation that is organic. But, don't use animal fats and bones. They decompose slowly and also attract dogs and other animals.

MULCH HAS VALUE, TOO

Mulch is a naturally good idea. Mulch preserves moisture, prevents weeds, improves soil condition, avoids erosion, cuts down disease problems, and adds organic matter to the soil. Next to composting, mulching is key number two for better gardening. The list of materials is practically endless, from leaves, chopped or ground brush, and twigs to straw, old hay, or pine needles. A mulch is any material that provides a protective covering for the soil including peat moss, sawdust, shavings, compost, and even gravel, sand, or stones. It also includes peanut hulls, ground bark, redwood chips, layers of newspapers covered with grass clippings, and whatever will decompose to add nutrients to the soil.

Newer black fabrics have become popular. I still prefer the black plastic sold for weed control in garden centers—covered with grass clippings to disguise it. Consider appearance and your goal, improving the soil and growing ground beneath any mulch. That means focus first on organic materials that break down and recycle into the good earth. Redwood chips and pine bark look nice but there are drawbacks with some materials. Wood chips, shavings, and cellulose materials draw nitrogen out of soil so you must compensate for that by adding higher nitrogen fertilizer to your garden ground.

Dry lawn clippings are fluffy when first spread. After a few rains, they tend to form a compact, thin layer. They do decompose and you can add more layers as the season progresses as you mow your lawn each week. The decomposition has its benefits of course by improving garden soil. By smothering weeds with mulch we also preserve soil nutrients and fertilizer we apply so they feed our desired herb plants.

Many natural gardeners prefer permanent mulching, which is basically like composting right in your garden. As mulch materials continue to rot, they give the added advantage of continued humus production. To plant more herbs, merely move the mulch. Frankly, I believe compost is the best mulch. As you create compost and humus, this decomposed organic matter provides an attractive and more useful mulch because it already contains some nutrients, but also through the composting process that generates high temperatures in compost piles, many weed seeds are killed.

KNOW YOUR ABC'S OF N–P–K

Expert gardeners focus on the big three, N-P-K, of plant nutrition. *N* stands for nitrogen, the key element for vegetative growth. This element promotes strong and healthy leaves, stalks, and stems. It is vital for all green leaf tissue. Nitrogen fosters development of proteins, the cell growth builders in plants. Without this essential element you'll have yellowed foliage and stunted growth. Plants become weak and susceptible to disease. Too much nitrogen also can cause problems, just as overeating does for people. Oversupply of nitrogen encourages excess leaf and stem growth at the expense of flower and fruit formation.

Too often, gardeners seem to believe that if X amount is good, they should double or triple that for better results. Not so! Another problem is using lawn fertilizer on flowers. Such formulas as 20-10-10 are high in nitrogen. It does its job, feeding grass well, but the higher nitrogen in lawn fertilizer formulations is wrong for herbs, flowers, and vegetables, too.

P is for phosphorus. It is vital for strong, prolific flower development, good fruit set, and seed production. Phosphorus also is required for proper development of plant sugars during the growth of plants, especially berry and tree fruits. Herbs need their share of this element, too. Lack of phosphorus is easily noted. Plants will be stunted and have a yellowed look. This may appear remarkably like nitrogen deficiency, but if there is a distinctive purplish color around edges of leaves it usually signifies phosphorus deficiency. Usually unseen is the retarded root development when phosphorus is lacking. Leaves may fall and plants may fail to flower. Check your herbs and other plants regularly for signs of problems. Plants will tell you when they are in trouble.

K stands for potassium. This basic ingredient promotes strong, healthy roots. Potassium or potash also helps in seed production. More important, K quickens maturity of plants and may help in disease resistance.

A deficiency in potassium is marked by yellowish mottling. In severe cases, foliage loss occurs and roots won't develop well.

Also, the formation of fruit is poor when potash is low. That's important if you have berry bushes, fruit trees, or flower shrubs that provide fall berries for attractive appearance and as food for birds.

Those are the essential fertilizer ingredients. They are available in varying ratios, as noted on the bag. They will always appear on fertilizer bags in the same order: N-P-K. Translated to numbers, 5-10-5 means a bag of fertilizer contains 5 percent nitrogen, 10 percent phosphorus, 5 percent potash. That means that a 100-pound bag of 5-10-5 will have 5 pounds of nitrogen, 10 pounds of phosphorus, and 5 pounds of potash. The rest is "carrier" material so the fertilizer spreads evenly. Higher analysis material is available, such as 20-20-20 and other concentrated formulations, which contain greater quantities of nutrients per bag.

Moderation Is Wise

Americans tend to overdo things. We eat a bit too much and tend to overdo our pesticide applications. Too often when we see a few insects we rush to blast them away with spray. Some novice gardeners seem to believe that if a little fertilizer works well, much more will give even better results. That's just not true. It is essential to read and heed directions on labels of pesticides and fertilizer materials, too. Read and heed is a good growing phrase to keep in mind!

Moderation works well in flower gardens, too. Focus on applying fertilizer in smaller amounts and on a regular basis to feed the plants what they need as they grow, mature, bloom, and set their roots and systems for the next year's growth.

Newer, slow-release and high-analysis fertilizers in various formulations are readily available nationwide. Schultz and Miracle-Gro are two popular national brands. Manufacturers also are putting more detailed product information plus tips and ideas on the packages and in free flyers at stores. That makes it easier to select the right nutrient combination plant food to meet your needs. Herbs will indeed respond to your TLC (tender loving care) with compost to improve soil, mulch to suppress weeds, and balanced fertilizer to feed their growing needs, especially if you harvest from the plants regularly.

If you still have questions about improving soil and choosing the right fertilizer, you have a valuable nearby asset, your county extension agent. These are talented, trained, and dedicated people who are state-federal employees. They are connected with the land grant colleges in their respective states and their services are paid for by your tax dollars. Tap into their knowledge whenever you need help or advice. They are a wonderful group of willing advisors with a vast store of local knowledge to share.

To be even more certain that you are feeding your Biblical herbs and other plants, as you should for your soil conditions, periodic soil tests are a good idea. Think of your soil as a bank. Your plants can only withdraw what the soil has in it for their needs. After that, you must make regular deposits of nutrients in your plant's account. As they grow, prosper, bloom, and set seeds or the leaves rebuild flowers within the bulbs for next year's blooms, plants naturally use up nutrients. It is up to you to maintain a reasonable balance of nutrients for your flowers.

Happily, the many modern fertilizer formulations for flowers, from bulbous flowers to other types, are readily available in convenient packages, labeled for various flowers. Read the package labels, consult the knowledgeable authorities at the garden departments of chain stores and local garden centers, and use what you need for the types of herbs, flowers, and plants you grow.

UNDERSTANDING SOIL PH

One more key point should be understood about nourishing plants. That is soil pH. Most herbs prefer a balanced soil but some like it a bit on the acidic side. Here's some background about that topic. You don't need to be a Ph.D. to understand it. Basically, pH is a method for measuring relative acidity and alkalinity in soil. Scientists have devised a simple chart, a scale ranging from 1 to 14, called the "pH scale." Acidic soil registers below the 7 mark, alkaline soil above that point. Most soils in eastern states tend to be more acid than western soils.

Acid soil is often called sour soil. Actually, many plants prefer soil slightly on the acidic side. That is because slightly acid soil helps break down both basic and minor nutrient elements in soil more easily so that they can be absorbed into plants more readily. To be picked up by plant roots, nutrients in the soil must be in solution.

Regular soil testing, with easy-to-use kits you can obtain from your local garden supplier, let you determine what your soil pH is. Be sure to check different areas. You may be surprised at the pH, and also the N-P-K, the nitrogen, phosphorus, and potassium difference in different parts of your home grounds, especially in moister areas. County extension agents, your reliable garden advisors in each county in America, also are helpful in providing tests and advice for improving your soil's condition and nutrition.

LEARN ABOUT LIME

Lime is not actually considered a fertilizer, but it is important for adequate plant performance. It sweetens overly acid soil. Bags of lime are inexpensive and widely used to improve soils for growing lawns. A few bags may be all you need to improve overly acid soil for your Biblical flower and herb garden areas, if necessary.

Flowers need phosphorus, but phosphorus is somewhat tricky. When soil pH is high, on the alkaline side, calcium deposits can lock up phosphorus into a calcium-phosphate compound, which makes it useless and unavailable to the plant. When you lower the pH, phosphorus is released. On the other hand, in acid soils, low on the pH scale, iron locks up the phosphorus. So, a soil pH balance of moderation must be maintained.

If some of this sounds difficult, you have an army of garden advisors and helpers available. Local garden centers and the garden areas of major chains often now have garden specialists who can give advice, answer questions, and guide your growing programs. Also check with longtime nurseries and private garden centers who are well established in the community. They are more likely to have extensive local knowledge of soil, growing conditions, and also sources for a much wider variety of plants than chainstore clerks can provide.

MORE GARDENING KNOW-HOW

In addition to the information in this and other chapters, you'll find many sources for free catalogs, good growing advice, websites, and links to a wealth of know-how listed at the end of this book. County Extension offices also have free gardening brochures on a variety of topics. They are listed in your phone book, and because they are state and federal employees, their advice is free. So is each state extension horticulturist, usually located at the State College of Agriculture. The U.S. Department of Agriculture also has many free and low-cost brochures and booklets.

Gardening today is easier and more fun and rewarding because so many people share an interest in making the good earth bloom and bear abundantly. Tap into the resources and sources available and you, too, can enjoy God's wonders in their deserved beauty and magnificence in your home grounds as well as in container gardens in cities.

Chapter Six

Grow Biblical Herbal Tastes Indoors

Container growing allows you to put a "garden" just about anywhere. Cement balconies on a high-rise building can become urban gardens to brighten lives and add the good taste of healthy herbs. Get creative! You also can shop flea markets and yard sales for old baskets, barrels, kettles, and buckets. Many flowers and herbs, Biblical ones and other types, too, will grow well in containers that are 8 to 10 inches in diameter and 8 or more inches deep. Taller plants and herbs with long taproots will need large, deep pots, tubs, or barrels. Match the container to the mature plant size.

As you plan your own container gardening keep these key points firmly in mind. Some plants prefer full sun while others can take partial shade. One of the advantages with containers is that you can move them around, especially if you use rolling bases to hold them.

Plants taller than 1 ½ times the height of the container may look unbalanced; therefore, as previously stated, match containers to mature plant size for best appearance. Another key thought for decor is to use groupings, according to the National Gardening Bureau. Think about one or two large pots with 2- to 3-foot-tall plants, one or two with 18-inch-high plants, and several with smaller plants. When placed together these herbs and flowers will provide a three-dimensional look to your mini-garden.

Coordinate pots, flowers, and leaves. Once you start playing with container plantings, you'll probably find many places for them. Place a big container in the middle of your garden. Try another amid ferns. Move pots around. Group pots on front steps, add them to deck rails, along paths, or in any odd spot. Put them on decks, under shrubs, just outside the garage door, or by the mailbox.

One common mistake with containers is choosing the wrong combination of plants. Since most herbs like sun and many can take some shade this isn't as big a problem as it is with flowers that require specific sunlight needs. Best bet is to have plants that like the same conditions in the same container, even if you can move them around. Another good design technique employs container plantings to soften the stark look of dull areas where no soil or garden area is available—such as side yards or walkways, decks, along garage walls, or cement areas bounded by chain-link fences. It's easy to build your own garden in these barren areas with containers. Line up window boxes or pots along a wall, for instance, placing them a few inches out. The containers can be all the same or a mix of any types you have.

A good growing medium is vital. In the past I often made up soil mix for containers using a combination of peat moss, vermiculite, and perlite. Today, you can simply purchase container garden soil mixes from garden centers. These professionally made mixtures are worthwhile. They provide ideal draining and growing conditions plus some have small amounts of fertilizer to get young plants started right. Simply add more nutrients with liquid fertilizers or pellets as needed during the growing season. Gardening gets easier all the time thanks to modern manufacturers who have tuned in to our needs and provided products that give us even better growing conditions.

As you begin container gardening remember that soil-less soil has been used by professional growers for years. They call it the ideal plant-growing medium. Soil-less soil is a well-balanced blend of peat moss, vermiculite, and perlite with plant nutrients added. Consider the advantages. Backyard gardeners must contend with bacteria, fungi, and insects. They must fight weeds and solve problems of dry soils and soggy clay ground. All those problems are avoidable with a sterilized, scientifically blended planting mix.

For container herb gardening with smaller herbs, simple clay pots work well. They have the advantage of providing needed drainage through the bottom hole and porous sides. However, they also dry out faster. Daily watering may be necessary. Plastic or metal containers tend to store heat and transmit it to the soil mix. That's useful in cool weather, but watch water needs carefully in summer. If you prefer more attractive containers, simply place clay pots inside more decorative planters.

Container plants need more water than those in backyard gardens because of their restricted growing habitat. Their roots cannot roam in search of moisture. They also are exposed to drying air on all sides from sun and heat radiated from building walls. Proper drainage is especially important also. Plants hate wet roots, and it's best not to water at night. Also be certain your containers provide for drainage and escape of excess water. Place several inches of coarse gravel in the bottom before adding soil mixture.

In pails, buckets, or baskets lined with heavy plastic bags to hold your growing medium, provide some holes whenever possible to let excess water escape. Plants don't like and in fact can't stand wet feet. It can stunt their growth, because plant roots need oxygen as well as moisture for proper growth and transportation of nutrients for the best plant growth.

It helps to test the soil moisture every day if possible. Poke a finger into the soil. If it remains dry, water well. Try the toothpick test to check moisture. Stick a toothpick into the soil. If it comes out with particles attached it has sufficient water. If it comes out dry, water the soil mix. Actually, it pays to give containers a soaking drink. Container gardens need deep watering to ensure that moisture gets to all parts of the soil mix. This encourages deeper and better rooting for sturdier plants. Avoid light sprinkling. That only promotes shallow surface rooting, which makes plants more susceptible to drought damage. Another good point is to use containers that have built-in watering systems. Insert nylon or cord wicks from a water tray or pans up into the draining holes of your containers to pull water into the lower levels of soil where plant roots can get it. Newer containers have special watering systems built in. When shopping for containers ask about these innovative new systems. Or, ask about the add-on watering methods now available for other types of conventional containers. As you water your plants, watch for their signals. In time you'll learn to spot their signs and tune in to container plant needs. Good for you and better for them.

Another key point is nutrition. Herbs may thrive better with poorer conditions than many flowers or vegetables but they do respond better to tender loving care. Plants need a balanced diet especially in containers where roots can't roam in search of food. Fortunately, many balanced fertilizer mixtures are available for container plant feeding. You have a choice of liquid fertilizer or prilled-type or slow-release pellets that feed plants over a longer period. Always follow the directions for the type you use. Too much fertilizer can be as harmful as too little, especially in container habitats.

Considering the container trend, plant breeders have focused on producing ideal flower and vegetable varieties for containers. Naturally, plants do require full sunlight, moisture, and nutrients, but new varieties offer spectacular results. Container culture allows the gardener to grow many other types of flowers, herbs, and even vegetables that are duplicate or relate to Biblical plants. Herbs, from sage and dill to other Biblical ones, do well in container gardens.

Watch your plants. Look for yellow or discolored leaves. Note drooping leaves. Yellowing or browning leaves may mean that more food should be applied. But be sure that you are not overwatering and drowning roots. That too will lead to yellow leaves. In time you'll get the feel for feeding your container gardens. Different plants have different nutrient needs. Each container may need different applications based on the amount of soil mix and number of plants in larger containers. Some herbs grow faster and have bigger appetites for plant food, especially if you are clipping and using them regularly. Keep that fact in mind and feed them well so they prosper and provide the bounty you want from them. Many fertilizers come with handy flyers that give tips and advice. Read and follow what the labels tell you to do for that brand, the types of plants that are listed, and amounts.

If you have doubts about feeding your container plants, ask local herb specialists or at the local garden center that provides your growing supplies. Or ask your neighbors. Gardeners love to share their knowledge, especially when they can show you examples of their container-growing success.

Today, in condo communities, many more people have adopted container gardens. Evidently they miss their backyard garden plots and many condo associations don't permit gardening around individual units. That is done by the landscape contractor. However, the use of attractive containers is often allowed. Ask at the condo association meeting. Growing together can be very rewarding.

Chapter Seven

Cultivate God's Beauty Everywhere

"A few seeds make a small harvest, but a lot of seeds make a big harvest." That's a friend's version of 2 Corinthians 9:6–11. I think of it when people say they would like to get involved with their church in a Biblical garden. The following pointers should help if you share the same inclination.

Look around at your home grounds or, with other members of your congregation, at the land or space available for a Biblical garden. Remember that most plants—herbs, flowers, and vegetables—like full sun, at least 6 to 8 hours per day where possible. What part of the church property can become a garden? If there isn't space, who in the congregation may have gardening space to "lend" for a Biblical garden?

The first basic step is to put together a Biblical garden team and together develop a vision or a mission for the garden. As you do, think of your garden as an attraction, not just for members of the congregation to enjoy but as a place that draws others from the community to your house of worship. Joseph Scott placed a Biblical garden on the front lawn of his church on a main road in Augusta, Maine. Every day during the growing season people stop to admire it and some admirers even began attending that church. Reverend Marsh Hudson-Knapp discovered that even when their Fair Haven Congregational Church garden was small it attracted people. When they added a water garden it became a meeting place for people to talk with one another while enjoying the beauty and serenity of that Biblical garden area.

Actually using a garden with an evangelism theme is very appropriate since Jesus himself focused on gardens and plants in his teachings. As you plan, carefully consider where is the best spot for your garden and what should be your first priority for plants. Ask yourself the following questions:

- Is the garden for enrichment and benefit of your congregation, young and old?
- Will your garden be an outreach program to attract others to "Grow with God"?
- Will you be trying to reach people who have never been to church?
- Will you be trying to reach people who have stopped going to church?
- Will you be trying to reach parents with children who are members or those who are not part of a church family?

Once you have developed a basic theme and plan, the next step is reaching out to others among your own congregation. In effect, you should plant the "seeds" of your garden with as many people as may be interested—seniors, parents, youngsters. Announcements at services, articles in newsletters, talks at circles and other groups, and a display at coffee houses all help present your Biblical garden idea and encourage support and more volunteers.

In the process, take a look around the church and ask yourself how someone new to the church would feel when they walked into your garden. Will it have that special welcoming appeal? Can it be maintained so it doesn't become overgrown, weedy, or an eyesore? Perhaps plan a special Sunday to talk about the garden project and explain how it can be a worthwhile and growing asset to your congregation and church or temple.

Next step is reaching out to the community. This also means reaching out for more gardeners to help with your project. Many churches and temples have found people willing to help with such gardens but who may not necessarily become members. In fact, Biblical gardens have served to unite people with the common interest of gardening but who retain their own different faiths and denominations. The gardens become an overall community beautification project. You can reach out with news articles in local media, plant sales, special mailings, personal meetings at other congregations, gardening seminars, and slide shows and similar efforts.

When people express interest, welcome them with open arms and hearts. They too can help all grow better with God and Biblical gardens. Involve children as soon as you can in planning and planting. Hands-on projects are popular and kids love to dig in the dirt. Planting flower bulbs, starting herb seeds in pots, and cultivating the garden each week together provide useful interactions. Kids can help look through colorful garden catalogs to pick and order seeds and plants. That whets their appetites. Consider setting up teams of children with adult leaders so they can plant and grow "their own" plants together. That makes a great multi-generational project. Benches in or near the gardens provide spots for children and visitors to sit and enjoy the gardens, meditate, and perhaps also become involved in the project.

Today there is much interest in gaining the attention of children away from TV and computer games. Think how your garden can be an asset. Plan to involve children in watering, weeding, and caring for their plants and keep telling them stories. Share Bible stories, life stories, and listen to their life stories! Children are eager to learn from those who listen to them and share lessons. Growing together can begin in the good earth. For example, each Sunday during the growing season tell a story with the children and bring in plants, blooms, herbs, vegetables, and fruit. Find the quotes about these plants in the Bible—as have been noted throughout the book—so you can relate the plants to the Holy Scriptures. Decorate the communion table and worship areas in church school rooms with plants and their products and use them as starters for exploring Biblical stories.

As you consider a Biblical garden, let your imagination soar. But, keep a pen and paper on hand to write down notes, plans, and even design the basic garden. As we know from moving furniture, it is far easier to move things on a paper plan than push furniture or dig and move plants in the garden. Even a small garden can get you started with a few herbs and lets you add more or expand with flowers, vegetables, trees, and fruits if you wish in the future. Remember that you can plant several areas. Herbs can be in one place, flowers mixed with them, or as a display elsewhere. Trees can be part of the landscape with flowering herbs around them. A Biblical garden can be fairly formal and all plants grouped together or with plants dispersed.

Do you want a memorial garden, dedicated to someone? Do you want meditation areas? Do you want sufficient herbs to harvest and use regularly? As you consider answers with your team members, be sure to find your USDA (United States Department of Agriculture) horticultural zone. Eliminate any plants that won't grow in your area unless you plan to grow them in containers and bring them indoors for the winter. Read about plants. Because most herbs and flowers prefer sun, place taller growing plants behind shorter ones. Be sure to give each sufficient growing room to mature without crowding. Also consider blooming and maturing times. Find plant maturity times in garden catalogs to include Biblical plants that will provide blooms and beauty as well as useful herbs over an extended growing season. That adds appeal to any garden.

Following are some basic precautions. Space may be scarce but don't plant a garden where heavy rain or snow falls off roofs. Even in winter that can damage or kill plants with weight, soil compaction, and excess water in the ground to rot roots. Find ways to divert roof runoff away from gardens. Consider soil improvement. If the area is backfill or poor soil, replacing it can be a first step to ensure proper growing conditions. Plan walkways so your gardens are accessible, including to persons with disabilities. Be aware too that kids and animals tend to wander freely at times. Perhaps an inconspicuous green woven wire fence about 2 feet high will deter neighborhood animals and young children.

Remember that "many hands make light work." Build a team of volunteers. The more you have the more you can all share the routine of weeding, feeding, and caring that comes with a Biblical garden after the fun of planning and planting is done. By starting small with a few plants you can probably entice more volunteers as you expand. Having a system of "dedicated" plants also encourages more people to become involved, especially donating money for such plants as memorials to relatives.

Don't forget signs. Biblical gardens need signs to identify the various plants and most important to provide that added dimension of Spiritual Knowledge that comes with them. They don't need to be fancy. It is important to have the proper names of plants, the Latin names, so people can know what to order if they wish to grow them at their homes. Naturally you should include the Scriptural references. Make your own signs on a computer and seal in plastic. Or find sources to make them for you. Some websites offer such services, including holders and markers of different types in which you can insert your own signs.

As you plan and begin activities, find out how others have successfully involved groups, built garden tender teams, and made Biblical gardens into valuable assets for the entire congregation. Reverend Marsh Hudson-Knapp also has other worthwhile thoughts that have worked well for his congregation. Draw and paint the plants and fruits of the gardens. A few snacks of Biblical foods, such as grapes, dates, cucumbers, melons, nuts, and apricots, can make the creating more fun and give a break to tell more stories.

"We are working on a calendar or booklet featuring the children's drawings, which we may publish here when we are done," Reverend Hudson-Knapp points out. "The children have done marvelous work and they feel proud and have come to believe in themselves as they create and are affirmed by adults."

Other projects based on their Biblical garden are worth considering too. Reverend Marsh suggests doing an art sale for missions with the children's artwork: "A few years ago our children did watercolors of Bible stories, displayed them during fellowship time and adults bought them, making contributions to Heifer Project."

Reverend Marsh continues, "We bring in food to share from our gardens during the summer and insist that people without gardens take home fresh vegetables. We hand out goodies on the way out of worship. Children can help make the deliveries. We teach adults and children to love the earth and to care for her as stewards. On one Sunday a year I put a bag of 'garbage' on a table and children help me sort out all of the recycling, compost, and reusable materials, leaving very little to dump back into our earth! We each do an environmental survey to evaluate how well we are doing in caring for our earth. Gardening and earth stewardship go hand in hand."

Your Biblical garden can help bring your community together, sometimes in unexpected ways. One year in Fair Haven, Vermont, a neighboring rabbi, Sol Goldberg, taught the Fair Haven Congregational Church how to celebrate the Sabbath, and through the month they made items to use on their "family tables." The children from each "family," made up of all kinds of groupings of people, hosted their table.

Other projects include drawing on the Biblical gardens for monthly church suppers during the season. Flowers and greenery grace the tables. Mint adds flavor to the tea. These are just a few ideas for integrating creative, edible, active, and earthy learning.

Another blessing occurred at the Fair Haven Congregational Church. People of all ages stop day after day at the gardens to watch things grow, to hunt out the frogs, to listen to the waterfall gurgle, and to feed the fish and hunt for the ones they have named. Young, old, and teens end up listening to and talking with one another. The water garden especially seems to draw people together as a community, as they do around the fountain in an Italian village. You too may find your garden becomes a gathering spot, just like gardens were in the time of the Bible for people of those times. Gardens have indeed been gathering places for eons.

Two questions inevitably arise about Biblical gardens, whether herb gardens or those that include many other Biblical plants. First, and it is a question that must be answered: How do you fund your Biblical garden? Second, How do you keep it well tended so it remains appealing and attractive?

Fortunately, a dedicated Biblical gardener, Stan Averbach of the Bucks County Jewish Congregation in Newtown, Pennsylvania, has worked diligently on these topics. In fact, he has dug up good information and with his permission, it is included in this chapter. His approach reveals his wonderful sense of humor, too. He willingly shares what he and his congregation have learned about funding a Biblical garden. Through God's grace they have been very successful and I hope you will find their ideas inspiring and useful. Following is Stan's "Sermon on the Amount":

> It is important to realize that the conditions for fundraising are different for each institution. So I cannot promise that our procedure will ensure success. But it is worth a try. Here goes:
>
> 1. Form a committee of interested people. Assign one individual who has administrative ability and one who has horticultural ability to those particular positions. The balance of the committee should be given assignments according to its needs. One person cannot do everything. It will end with burn-out. Make sure the committee will "walk-the-walk" and not only "talk-the-talk."
>
> 2. Measure out the area that can be planted, make a plan, and present it to the board and the clergy of the congregation. The presentation should list the benefits to the congregation and the community. It should also state how the committee intends to finance the garden with little or no cost to the congregation. Arrange a contingent loan from the congregation as a backup, not

to be used except in case of a temporary financial shortage. This approach should spell success.

3. Have your newsletter make an announcement as to your intentions. Have your clergy present a sermon regarding the spiritual and aesthetic aspects of such a garden and the beauty of memorializing family members and as honorarials for special people and special occasions.

4. Set up prices for plants, shrubs, trees, planters, benches, arbors, walks, etc. If you have a spiritual number in your religious denomination that is meaningful to your congregation, use that as a base or multiples of it to set your prices. For example, in Judaism the Hebrew letters that stand for 18 also is the Hebrew word for life. It is a common tradition to give charity of $18 and multiples thereto. We believe that living memorials and honorarials seem to attract many people. They are contributing to a living thing that they can visit and watch it grow and blossom or bear fruit. They can see the plaque at the base and reminisce. Also they can find a place for quiet meditation and spiritual thoughts.

5. You can get paver kits for walks or garden borders. They are 1 foot square, round, or other shapes at chain stores. Have special days where children or adults can make impressions of their hands and inscribe their names and birth dates. I would suggest that, after 2 weeks of curing, you give it 2 coats of clear waterproof paint to minimize any cracking due to expanding water in the impressions.

6. If you have a contribution program for various funds, institute one for a Biblical Garden Fund. We find that people have been generous with contributions but not with physical labor since both family members are too busy. This may be important in order to be able to buy seeds, plants, supplies, and professional help to prepare the garden for seasonal changes.

7. If you have Good Deed Days where the congregation reaches out to the community to perform various

projects, encourage some congregants to do some work in the garden.

8. If you are in a small town, have a large plaque listing companies that have contributed to the garden's existence. It can be in supplies, funds, or both.

9. Encourage continued interest in the garden through the education of the children and adults.

10. Come up with your own ideas and tell the Biblical garden community what worked for you.

As the Babylonian Talmud states, I believe, in Tractate 23, "I shall plant a tree. I shall plant a tree for my children." Yours in brotherhood and love, Stan Averbach.

And to be of even more help to those who want to develop a Biblical garden at their church, synagogue, school, or anywhere, he offers his wit and wisdom by phone. As Stan says, "If there are any more questions, I can be reached at (215) 946-3859."

Also at the end of this book, I've listed my favorite Biblical gardens in the United States as well as my list of favorite gardening websites. These are the sources of talented people at many horticultural and mail-order companies, manufacturing concerns, and organizations that have guided me for decades as a garden writer. On those sites you'll find a bounty of basic and also advanced gardening information. Some sites also have toll-free numbers to call for answers to specific information. Finally, I have listed a plentitude of free garden catalogs. Most include herbs. Some specialize in them. Look over the list and request a few that seem to fit your needs. Today many catalogs have wonderful descriptions of plants. Some now include useful growing tips, which can also be found on the seed packages and flyers that mail-order companies send with orders.

The following three are excellent sites to start you off in your Biblical garden research. Steve Hallowell adds to his site regularly as an enthusiastic Biblical gardener, www.gardenprojectonline.com. I recommend www.biblicalgardens.com/index.htm, launched by Shirley Sidell who was instrumental in developing the marvelous Biblical garden at her temple. This site has lots of information, including a searchable database and also has a store to order some Bible garden seed mixes to get you started quickly. Two other sites that reference Biblical plants include www.seedman.com/bible.html and www.newadvent.org/cathen/12149a.htm.

One church that has one of the most informative sites is the First Congregational Church in Fair Haven, Vermont, put together by Reverend Marsh Hudson-Knapp. Visit www.sover.net/~hkfamily/index.html. He has written several very useful folders and literature focused on how he has developed that garden. Details are on the website along with a wide list of Biblical plants. Other websites will provide you with other plant lists as Biblical gardens grow them around the country. If you don't have a computer or Internet access, check with your local library. Most do and many offer classes or help so you can explore these Biblical gardens and add to your store of plant knowledge.

One of America's leading horticultural therapy gardeners, Charles Sourby, had some very worthwhile points to make about plants of the Bible as they relate to healing and therapy.

"Plants mentioned in the Bible are a living link between us and the people of those hallowed times," Sourby believes. "The humble onion and leek, the dill, mint, and rue of our present days were also part of the everyday life of the people of the Bible. Actually, the Bible is a book of gardens from Eden to Gethsemane," Sourby explains. "It is with the garden that the book of Genesis begins and with a vision of trees bearing fruit that Revelation ends."

As we think about Biblical plants it is appropriate to remember that gardens have always seemed essential to a person's life. In Biblical times they were respected gathering places. The Scriptures tell us that Jesus and his disciples often went to gardens to rest, meditate, and pray. As Charles Sourby reminds us, it was in the Garden of Gethsemane that Jesus made his great decision, and it was in the garden of Joseph of Arimathea that the body of

Jesus was laid. It also was in that garden where the weeping Mary mistook the risen Savior for the gardener.

With many thanks to horticultural therapist and Biblical gardener Charles Sourby, I have included some of his key thoughts here because they deserve serious contemplation. He notes in the courses he teaches at the New York Botanical Garden and elsewhere that there is a spiritual nature to plants and that they have played a deep and significant role in the Bible. Plants are more than a background, he explains. They are part of the drama and panorama of life.

"Plants clothe the land of Palestine in a rich and luxuriant verdure; they clothe the thoughts and speech of the prophets and leaders of the land with rich poetry and imagery. They are symbols through which spiritual ideas can be presented. For us, the symbols can live in growing green as we plant a Bible garden," Sourby explains.

For example, he notes that many trees and shrubs mentioned in the Bible are excellent shade and specimen trees and ornamental shrubs for the home grounds, hospital grounds, or churchyard. As you read about herbs to grow in this book, be especially aware of Sourby's recommendations. The herbs of the Bible can be planted in flower beds or used as edgings and borders. Other plants such as cucumber and melons have, as we've seen, meaningful Scriptural roots. As you plan your own home garden or work with others at your house of worship, heed these other worthwhile thoughts from Charles Sourby. He emphasizes that "In a garden we find a space where we can feel both removed from our daily cares and connected to nature's grace. By planting herbs and tending flowers, foliage, and vegetables our link with the earth which has sustained us and nourished us since the beginning of human history, is reestablished.

"For many of us, it is in the solitude and safety of the garden that we experience the deepest sense of peace. We associate green with nature's cycles as the emblem of new growth. Seats and benches in a garden allow us to share a moment or a meal with our dearest ones. In our everyday lives we may accept change grudgingly but in the garden we nurture and celebrate its stunning beauty," Sourby says. I agree wholeheartedly.

Gardens offer us all a mystique that must be experienced. Soul searching is a healthy pursuit. That's why a seat or a bench in a quiet garden area provides such a useful meditation spot. The aroma of ancient herbs and flowers remind one of their history to antiquity. Crumbling a few herb leaves can bring back fond memories.

Chapter Eight

Biblical Gardens to Visit

A wealth of great growing knowledge is available at Biblical gardens around the United States—and from their gardeners, people of vibrant abiding faith. When I first began researching Biblical gardens in 1980 I could find only a handful of gardens based on replies to my letters and phone calls nationwide. Happily, the idea of planting Biblical gardens has taken root and grown magnificently.

I've been able to visit with hundreds of wonderful people who have been tending marvelous Biblical gardens since that time, some in countries around the world, too. And many a mouse-click away on the Internet! Some of the gardens they're involved with are amazingly extensive, with more than 100 different plants growing in gracefully designed plantscapes. Others are more modest, but equally well researched based on Scriptural passages. None are specifically herb gardens although herbs play a part of their magnificence. With appreciation to all who have planned and now tend and faithfully care for these Biblical gardens and share their knowledge, the following is a brief tour of them with some personal comments from the founders, today's dedicated tenders, curators, and directors. Remember, these are marvelous people who love to share their love of God and their Biblical gardens. I trust you will be in touch with some of them.

CATHEDRAL CHURCH OF ST. JOHN THE DIVINE BIBLICAL GARDEN

Cathedral Church of St. John the Divine
1047 Amsterdam Avenue
New York, NY 10025
(212) 678-6866
www.stjohndivine.org

This Biblical garden is one of the most impressive as it grows at the side of one of the largest cathedrals in America. Located on the 13-acre complex in which the cathedral stands on Morningside Heights in New York City, it contains numerous species that existed in the Holy Land. Thousands of visitors marvel at its beauty and learn about the plants each year. At the time of Christ most gardens were primarily functional, providing food and fruit for sustenance of the people. Gardens also were grown for their beauty and served as sanctuaries where people could find refuge, tranquility, and a place for meditation and prayer. With the assistance of C. Powers Taylor of Rosedale Nurseries of Hawthorne, New York, this garden was lovingly planned and planted. It was begun in 1973. This original garden was restored by Mrs. Alexander O. Victor in 1988. The guiding force and main support for the garden is the Cathedral Guild, which is headed by Mrs. Henry "Betty" Thompson. "Every flower, herb, tree, and shrub that grows there is related to plants that existed in the Holy Land 2,000 or more years ago," Mrs. Thompson points out. Visitors enter the redesigned garden through a 10-foot-tall wooden lynch gate, a roofed portal that is common to churchyards. They are greeted, in appropriate seasons, by a dazzling profusion of flowers, shrubs, and trees, including crocus, iris, sage, thistle, lily, and almond, all of which can be traced to their heritage in the Holy Land. A bluestone walk in the shape of a cross bisects the garden.

"I saw many people throughout the day sitting in silence in the garden," Reverend Jay Wegman said in a recent interview for *The New York Times*. "This garden is a respite." The grounds of the world's largest Gothic cathedral are open daily from 7:00 A.M. to sunset. Major viewing months are May to October. Special group tours can be arranged.

CHURCH OF THE HOLY SPIRIT BIBLICAL GARDEN

Church of the Holy Spirit
Biblical Gardener—Page McMahan
9 Morgan's Way
Orleans, MA 02653
(508) 255-0831
www.holyspiritorleans.org
www.diomass.org

For years volunteers have been dedicating their efforts to this marvelous small but attractive garden that has more than 40 different plants from aloe and anemone to thistle and even wheat. Among the flowers of note in Page McMahan's list are anemone, saffron crocus, daylily, flax, iris, Madonna lily, mustard, narcissus, nasturtium, flowering quince, Provence rose and climbing rose, the Star of Bethlehem, and globe thistle. Located on Cape Cod, this garden bids welcome to the multitudes who visit that scenic part of America's seashore.

A Biblical garden, mostly of herbs, grew at this church for years. Recently, as part of a major parish building project, the dedicated Biblical gardeners, led by Page McMahan, expanded the garden. Although they do not have large specimen plants, they have included a wide range of Biblical plants to present a truly representative collection. For additional information from this dedicated and sharing gardener, call the above-mentioned phone number.

CHURCH OF THE WAYFARER BIBLICAL GARDEN

Church of the Wayfarer
Corner of Lincoln & 7th
Carmel-by-the-Sea, CA 93921
(831) 624-3550
www.churchofthewayfarer.com

This marvelous and deeply rooted garden began with Mrs. George Beardsley's inspiration and generosity about 1940. Many people have contributed to its care and growth. The village of Carmel and its Methodist church, the Church of the Wayfarer, have a long history in California. Carmelite friars marveled at the "Garden Land" a translation of the Hebrew "carmel" which they saw as a mirror of Mt. Carmel of the Holy Land. The Church Biblical garden is named "The Master's Garden." From the church door, one can enter the garden to enjoy the flowers, herbs, trees, and other Biblical plants. A detailed booklet with Scriptural passages and information about all plants is available and one phrase stands out, "For flowers are the poetry of Christ." A most fitting tribute to the faith of many who have faithfully tended this beautiful garden throughout the decades. The United Methodist Church designates one Sunday each year as a "Festival of God's Creation" or "Earth Sabbath." The Biblical garden is one way they have chosen to enable congregations to celebrate God's gracious work in creating the earth and all living things that dwell upon it.

CONCORDIA LUTHERAN CHURCH BIBLICAL GARDEN

Concordia Lutheran Church
40 Pitkin Street
Manchester, CT 06040
(860) 649-5311
www.macc-ct.org/member_churches.htm

An extensive garden with hundreds of flowers graces the grounds of Concordia Lutheran Church, according to enthusiastic gardener Jan McGarity. Jan describes this as a memorial garden that has some Biblical plants. A "Pathway of Faith" is engraved with 24-by-24-inch stones, which have names of the donor supporters of the garden. The garden includes an outdoor altar for services, and baptisms have also been held here. Jan McGarity is justly proud of their garden and shares information and ideas. She can be reached by phone at (860) 872-6266, or via e-mail at Bobmcgar@aol.com.

CONGREGATION B'NAI SHALOM BIBLICAL GARDEN

B'nai Shalom Synagogue
74 Eckley Lane
Walnut Creek, CA 94596
(925) 934-9446
www.biblicalgardens.com
www.jfed.org

Avid veteran gardener Shirley Pinchev Sidell is called a visionary by those who know her. She not only has helped create a marvelous Biblical garden at her temple, but also has expanded her growing and sharing horizons around the country. The synagogue is located on an 8-acre hilltop in Walnut Creek, California. She conceived the Biblical garden during a trip to the Holy Land several years ago. Most of the plants she saw were familiar because "we share similar growing conditions and many of the same plants are found in my garden," she notes.

At that time, Shirley was chair of the landscape committee. She wanted to begin a Biblical garden but could not find adequate resources at that time. As she continued research she began assembling a database, which led to another productive direction. She founded biblicalgardens.com with a vision to help "build a Bible garden in every home, church, synagogue, or temple using the Internet," she explains. In the past 2 years gardening friends have donated millions of seeds; others have spent hours cleaning and sorting seeds for her free-seed offer via her website.

The B'nai Shalom Synagogue Biblical garden was scheduled to be dedicated on September 16, 2001, the evening just before the start of the Jewish High Holy Days. In keeping with her focus on accuracy and shared information, all plants are well identified with descriptions and botanical as well as Scriptural references. Seed kits also are available for sale. You can obtain information via www.jfed.org/bnaishalom.

FAIR HAVEN BIBLICAL GARDEN

First Congregational Church
2 North Park Place
Fair Haven, VT 05743
(802) 265-8605
www.sover.net/~hkfamily

In 1981 Reverend Marsh Hudson-Knapp visited Israel where he developed an interest in the plants of the Bible. A Biblical garden at the First Congregational Church "sprouted" in his mind and it thrives today thanks to his wife, Cindy, and other dedicated church members. They have nurtured the gardens for more than 20 years and now enjoy a wide range of flowers, herbs, trees, and other representative Biblical plants. By the turn of the millennium there were 73 different plants from flowers and herbs to vegetables, fruits, and trees, as well as food crop plants in that marvelous garden.

"All through the Bible, trees and flowers, fruits and vegetables play prominent parts," Reverend Hudson-Knapp says. "Every year our Jewish brothers and sisters celebrate their deliverance by observing the Feast of Passover. One part of the seder meal involves eating bitter herbs, a reminder of how bitter life was for our ancestors when they were slaves." Dandelions have been identified as one of those bitter herbs and have a certain blooming beauty that qualifies them as a flower to be considered for inclusion in this and other Biblical flower gardens.

The Fair Haven garden blooms in season with crocus, hyacinths, tulips, daffodils, narcissus, and iris, too. "The writer of Ecclesiastes also speaks of the beauty of wisdom, comparing it with the rose of Jerico. Our rugosa rose is a modern version of that simple but beautiful flower," Reverend Hudson-Knapp explains. Jesus drew his follower's attention to the plants that bloomed abundantly around him as signs of God's abiding care. Focusing on one of the teachings, "consider the lilies of the field," Reverend Hudson-Knapp has planted an array of flowers that could have been such plants. Shasta daisies, crown anemones, ranunculus or wind flowers, chrysanthemums, delphinium, hibiscus, and lupines all bloom gloriously in season.

The Fair Haven website offers much more wisdom from the Scriptures and also contains links to other Biblical gardens and gardeners in America and around the world. The website features a chart that includes pictures of Biblical plants, their common and Latin names, their significance and meaning, and equally important, the appropriate Scriptural passages. The Hudson-Knapps' goal, like that of many other Biblical gardeners, is to encourage many more people of all faiths and denominations to plant and nurture Biblical gardens.

Sometimes websites only take you to that facility's home address and don't link to their Biblical garden, Marsh reminded me. On his exceptional website he has links to several key Biblical gardens. Shir Ami is at www.sover.net/~hkfamily/Pages/Shir%20Ami.html. St. John's in Worthington, Ohio, is at www.sover.net/~hkfamily/Pages/Worthington.html, and the new Kavanaugh Gardens are at www.sover.net/~hkfamily/Pages/Kavanaugh.html.

FIRST PRESBYTERIAN CHURCH BIBLICAL GARDEN

First Presbyterian Church
4815 Franklin Pike
Nashville, TN 37220
(615) 383-1815
www.klis.com/scove/041400.htm
www.fpcnashville.org

Grace Westlake moved to Fairfield Glade, Tennessee, in 1991 and one of the first people she met was Kay Greene, an herb gardener and member of the Community Church, affiliated with the Presbyterian and United Church of Christ. Their interest in Biblical plants led naturally to the creation of a garden. Today, the garden includes herbs in a wheel-shaped bed, a triad garden with seven beds including flowers that bloom in spring, yellow flag iris and a variety of hyacinths, and vegetables added for summer. A cross is the focal point of the garden, forming four beds with biblical flowers from Madonna lilies to anemones with some blue flax, mallow, loosestrife, and tulips. A ladder garden with other flowers from narcissi to sweet flag, larkspur, and lilies, bloom in that special spot.

HIGHLANDS PRESBYTERIAN CHURCH BIBLICAL GARDEN

Highlands Presbyterian Church
1001 NE 16th Avenue
Gainesville, FL 32601
(352) 376-2440
www.gnv.fdt.net/~hpc/garden/

Visitors can sit by a bubbling fountain and look upon many of the plants and trees mentioned throughout the Bible at the Highlands Presbyterian Church Biblical Garden in Gainesville,

Florida. Their website even offers a musical interlude, "In the Garden," as the visitor contemplates God's majesty in the quietness and seclusion of God's creation.

MAGNOLIA PLANTATION BIBLICAL GARDEN

Magnolia Plantation and Gardens
3550 Ashley River Road
Charleston, SC 29414
(800) 367-3517
www.magnoliaplantation.com/gardens/index.html

One of the most dramatic Biblical gardens in America has been created by Drayton Hastie, owner of Magnolia Plantation and Gardens in Charleston, South Carolina. That outstanding plantation dates to 1670 and has been meticulously maintained with traditional Southern gardens by 11 family generations. It includes one of America's most dramatic Biblical gardens, which was created following the publication of my first book, *Your Biblical Garden,* in 1981, according to a plaque at the site.

As visitors explore the plants and flowers of the New Testament, which surround the cross, and in the beds of the cross representing the 12 disciples, the visual impact that greets them is extraordinary. So, too, is the visual picture as one scans the gardens of the Old Testament area surrounding the Star of David, which commemorate the 12 tribes of Israel. "We at Magnolia Plantation and Gardens are delighted that you are writing another book on Biblical Gardens," says Taylor Drayton Nelson, grandson of the present owner. "We regularly receive calls from people wanting information about our garden so they can develop one at their church or in their community and are always pleased to be of help and can provide sources for Biblical plants too" he says.

OJAI PRESBYTERIAN CHURCH BIBLICAL GARDEN

Ojai Presbyterian Church
304 North Foothill Road
Ojai, CA 93023
(805) 646-1437
www.ojaipc.org

The Ojai, California, Biblical garden attracts visitors from around the state. This garden sprung from a Biblical flower show arranged by Mary Hunt in 1954. Today when visitors enter the garden they receive a map of the grounds with names of all the plants and their marked locations. At the base of each specimen is a sign giving the plant name in Latin and the Bible references. Four Bibles were used: the King James Version, the Jerusalem Bible, the Revised Standard Version, and the New English Bible. A visit to this Biblical garden helps guests understand Scriptural passages as they relate to actual identification of the plants. Tours of the garden can be arranged by calling the phone number listed above.

PARADISE VALLEY UNITED METHODIST CHURCH BIBLICAL GARDEN

Paradise Valley United Methodist Church
4455 E. Lincoln Drive
Paradise Valley, AZ 85253
(602) 840-8360
www.pvumc.org/about/biblicalgarden.html

The idea for a Biblical garden at Paradise Valley United Methodist Church grew from the similarity of Palestine's climate to that of Arizona, according to Bob Morgan, the tender of that delightful spot. "The garden is located behind the original chapel and includes flowers as well as other Biblical plants. Anemone or 'lily of the field' produce a mass of red blooms in the garden around Easter each year here in Arizona, just as they cover the plains near the Sea of Galilee," Morgan points out. Other key plants include olive, papyrus, myrtle, olive, palm, pine, and pomegranate.

PRINCE OF PEACE LUTHERAN CHURCH GARDEN

>Prince of Peace Lutheran Church
>209 Eastern Avenue
>Augusta, ME 04330
>(207) 621-1768
>www.poplink.org

Another well-focused Biblical garden began several years ago and now provides a visual and aromatic perspective of the flowers, herbs, and other plants mentioned in the Bible. Joseph Scott, a member of the church, took on the research and design. Begun in 1997, the garden is a 20-by-24-foot plot on the front lawn. Today the garden includes 40 different plants native to the Holy Land. Because some species were not available, Scott had to seek alternates. "When a substitute was made, the same genus and family was used and looks very much like the one found in the Holy Land," Scott notes. "This garden is designed to look as if you cut a chunk out of a field in the Holy Land and transplanted it to Maine."

RODEF SHALOM BIBLICAL BOTANICAL GARDEN

>Rodef Shalom
>4905 5th Avenue
>Pittsburgh, PA 15213
>(412) 621-6566
>www.rodefshalom.org/Garden/initial.html

One of the largest Biblical gardens in America, a third of an acre, is Rodef Shalom Biblical Botanical Garden in Pittsburg. This dramatic garden has more than 100 varieties of flora. This garden is reminiscent of the Holy Land, including a cascading waterfall, a desert, and a bubbling stream that represents the Jordan River. Director Irene Jacob explains that this extensive garden was begun in 1986 and presents the most complete array of Biblical plants in North America.

The Rodef Shalom garden includes foods—wheat, barley, millet, and many herbs along with olives, dates, pomegranates, figs, and sycamores. Lotus, papyrus, rushes, and water lilies are on display. Details about the garden and program are available by calling Irene Jacob at the telephone number listed above.

ST. GREGORY'S EPISCOPAL CHURCH BIBLICAL GARDEN

St. Gregory's Episcopal Church
6201 E. Willow
Long Beach, CA 90815
(562) 430-1311
www.stgregoryschurch.com

This Biblical garden may be one of the more complete among those in the United States, according to cofounder Betty Clement, because it has 86 plants of the Bible growing well and providing dramatic, meaningful displays to parish members and visitors alike. A retired school principal, Betty has methodically researched Biblical plants for authenticity. The garden is artistically designed with a piled rock fountain in one corner, crushed rock paths stated with four impressive trees, a braided ficus, a date palm, olive, and pomegranate. Narcissus, poppies, and tulips are among the dramatic flowers. Information is available from Betty at (562) 421-4918. She also does slide shows for groups and her joking motto is "Have slides, will travel."

ST. JAMES LUTHERAN CHURCH
BIBLICAL GARDEN

St. James Lutheran Church
110 Phoenetia Avenue
Coral Gables, FL 33143
(305) 443-0014
www.st-james-church.org

During the past few years a large number of Biblical plants and shrubs have been successfully established in the Garden of Our Lord at this church. It has been developed as a community- and nation-wide project.

According to their informative website, "The Song of Solomon lauds the costly spikenard and aloes, to be found along the paths of this garden as are hyssop, mint, and cinnamon. The small bushy myrtle, used by Hebrews for perfume and spices, the lily and the bulrushes such as were used for the ark in which the infant Moses was found lying along the Nile, as well as willows are included in the garden." In this marvelous garden are flowers grown from seeds sent directly from the Kidron Valley in the Holy Land.

ST. JOHN'S EPISCOPAL CHURCH
BIBLICAL GARDEN

St. John's Episcopal Church
P.O. Box 2088
Norman, OK 73070
(405) 321-3020
www.episcopalnorman.org/outreach.htm#garden

This beautiful garden graces the church grounds at St. John's Episcopal Church. The Biblical Garden Guild was created to plan, establish, and maintain a collection of plants abundant at the time of the Old Testament and has generated many requests for information from organizations wanting to establish their own

Biblical gardens, according to the director, Betty Burns. She invites all church members to volunteer for a variety of tasks, from planting and weeding to ordering plants. She also kindly answers questions at the above telephone number. More details are on the website.

SHIR AMI BIBLICAL GARDEN

Bucks County Jewish Congregation
101 Richboro Road
Newtown, PA 18940
(215) 946-3859
www.uahc.org/congs/pa/shirami

The Shir Ami Biblical Garden has sprouted well, nursed along by Jules Hyman and Stan Averbach, Shir Ami Bucks County Jewish Congregation, located in Newtown, Pennsylvania, and is now 30 by 650 feet long! This garden began in 1999. Their regular garden is called "Shomrei Gan," which means "Keepers of the Garden." It is 30 by 400 feet, laid out with a winding path of pea-sized stones over heavy plastic sheeting and sides to minimize weeds. The path symbolizes the path of life that has many turns and can move in mysterious ways. The plants, shrubs, and trees are planted in the hollows of the path, and, where possible, in Biblical order starting with Genesis. A fig grove, an apricot grove, and an apple grove are included, with a pine tree, walnut tree, and oak tree. This garden also contains wheat, barley, sorghum, millet, flax, and cotton in earth box planters to control growth. By 2001 there were 56 trees and shrubs planted and 4 grapevines.

The children's garden is adjacent to the synagogue school and is 8 by 100 feet. Flowers, vegetables, and herbs are planted in earth boxes and large planters are placed in pairs parallel to the walkway, according to Stan Averbach. Stan Averbach (Poppopstan@ aol.com) loves to exchange Biblical garden knowledge, tips, and advice.

Photos can be seen at the Fair Haven website via www.sover.net/~hkfamily/Pages/Shir%20Ami.html.

STRYBING ARBORETUM BIBLICAL GARDEN

Strybing Arboretum
9th Avenue at Lincoln Way
San Francisco, CA 94122
(415) 661-1316
www.strybing.org

This carefully documented collection includes plants mentioned in the Bible or thought to be growing at the eastern end of the Mediterranean Sea during Biblical times. Plants then were a vital part of everyday life for food, medicine, fragrance, seasoning, and clothing. The Strybing Biblical Garden features fragrant and flowering plants that have been attractively landscaped to suggest the potential use of these in home landscapes. That's a plus for all who want to better visualize how a Biblical garden should be designed for their house of worship or especially their own home landscapes. Many shrubs and herbs are included and worth seeing. Featured flowers, including rockrose, violet, lavender, sage, anemone, crocus, autumn crocus, cyclamen, hyacinth, iris, lily, narcissus, Star of Bethlehem, sternbergia, and tulip, make this one of the best botanical gardens in America. Enjoy a "virtual walk" by logging on to their website, a most colorful and informative site for all gardeners.

TEMPLE BETH-EL BIBLICAL GARDEN

Temple Beth-El
70 Orchard Avenue
Providence, RI 02902
(401) 331-6070
www.temple-beth-el.org

A highlight of this garden is a shaped pyracantha or firethorn in the form of the 7-branched candelabra known as the menorah, according to founder and first president of the Eden Garden Club of Temple Beth-El, Mrs. David C. Adelman.

"Because of our climate or the exposed location, or because the plant I wanted for biblical authenticity happened to be an annual or I could not obtain the exact materials that I sought to achieve biblical continuity of special interest, I used individual plants with symbolic backgrounds," Mrs. Adelman noted in her comments for the garden's information folder. That makes sense to many who feel frustrated trying to obtain Biblical plants. As other gardeners have discovered, it is better to select closely related plants that will grow, thrive, and perform well, rather than risk poor growth and performance. Presently Martha Finger has been overseeing the garden with other active members of the Eden Garden Club including president Frances Sadler, Janet Freedman, and Betty Adler.

TEMPLE BETH SHALOM BIBLICAL GARDEN

Jewish Community Center of Sun City
12202 101st Avenue
Sun City, AZ 68535
(602) 977-3240 and (602) 972-4593
www.goodnet.com/~tbsaz/body_index.html

A Biblical garden has been thriving since 1988 at Temple Beth Shalom and the Jewish Community Center of Sun City, Arizona. In 1986 the late Rabbi Bernard Kligfeld remarked, "Wouldn't it be nice to have a Biblical garden?" They now have—designed, planted, and maintained by Colonel Hy Mandell, USA Rtd. Each year since the original planting, on Tu B'Shevat, a new planting is made with appropriate ceremonies, Hy Mandell explains. "To date 40 plus trees, bushes, and other plants are growing on a ¾ acre garden. A sample includes palm, acacia, black and white figs, citron, almond, olive, pomegranates, carob, walnut, allepo pine, and cypress. Herbs include sage, aloe, rosemary, and magnificent frankincense," he points out. "The guiding light in our planting is that what we put into the ground must be as identified in the Torah, commonly referred to as the Old Testament"

Colonel Mandell identified Mountain States Wholesale Nursery, P.O. Box 2500, Litchfield Park, AZ 85340 and Sun City Nursery, 9715 W. Peoria Ave., Peoria, AZ 85380 as useful centers for finding Biblical plants and herbs.

TEMPLE SINAI BIBLICAL GARDEN (THE EDWARD E. KAHN MEMORIAL BIBLICAL GARDEN)

Temple Sinai
11620 Warwick Road
Newport News, VA 23601
(757) 596-8352
www.ujcvp.org/temple_sinai/bulletin04-00.html

The Edward E. Kahn Memorial Biblical Garden is replete with dozens of glorious Biblical flowers and plants. As Curator Lee Kahn Goldfarb explains, "From the Garden of Eden in Genesis to the tree of life in Revelation, almost every book of the Bible contains references to flowering plants. The idea of this Biblical garden emerged as something that might be of interest to members and friends of Temple and was also seen as a valuable and instructive tool for children."

From crocus and narcissus to the pale blue flowers of flax, to the anemone, iris, and tulip identified with Scriptures, to the blooming almond trees, this garden encompasses a wide range of plants. In addition, hibiscus, larkspur, lavender, water lily, lupine, marigolds, Phoenician rose, globe thistle, and violets are included. Some botanical purists may not agree with some plants on their list, but no doubt extensive research has been done. The Temple Sinai website contains their full list of plants. It is worth accessing for historical as well as your own garden planning purposes.

WARSAW BIBLICAL GARDENS

Warsaw Biblical Gardens
P.O. Box 1223
Warsaw, IN 46580
www.warsawbiblicalgardens.org

Another wonderful site is Warsaw Biblical Gardens, just off busy State Road 15 in Warsaw, Indiana. It is filled with more than 100 types of plants mentioned in the Bible.

These lovely gardens were created on the site of a former scrapyard by the owners, the Levin family, who had an idea for the Biblical gardens. Their website is being updated.

You'll find a gathering area, a sitting area, forest, meadow, some food and crop growing spots, a desert and orchard, and winding brook for water plants. It is a slice of Holy Land habitat in which Biblical plants grow, tracing their roots to the Scriptures.

The Warsaw Biblical Gardens serve as a public research center for the study of plants of the Bible. It is educational, inspirational, and accessible to persons of all ages and faiths as a community garden. As such, it stands as positive evidence of what a family and their community can do to provide an uplifting, educational opportunity to help people of diverse backgrounds grow better together.

OTHER BIBLICAL GARDENS ARE SPROUTING

With the new millennium, I'm pleased that there are other Biblical gardens growing, being planted and planned. Good friend and exceptional Biblical gardener Reverend Marsh Hudson-Knapp advises that a Garden Club at Liberty Presbyterian church, a country church about 12 miles north of downtown Columbus, Ohio, has started work on a Biblical garden. Marvin Languis has been gathering information for their designer Tom Wood to use.

Nearby, Reverend Arthur Hadley, rector of St. John's Episcopal Church in Worthington, Ohio, reports that they planted

1,000 bulbs in two plots of wheat to be the lilies of the field. They also have ordered rosemary, fig, apricot, grape, and more. They are raising from seed or cuttings fig, olive, pomegranate, and date palm as Sunday school projects. They plan a pergola with seating for private meditation in the garden. Some photos can be seen via the Fair Haven website organized by Reverend Marsh Hudson-Knapp: www.sover.net/~hkfamily/Pages/Worthington. html. Coincidentally the Worthington Church is only a few blocks from St. Michael's Church and 5 miles from Liberty Presbyterian, both of which have Biblical gardens now.

Reportedly another garden is being planned at the First Congregational Church in Winter Park, Florida. Reverend Ken Crossman, a retired United Methodist pastor, is working with pastor Bryan Fulwider to plan a garden to launch 2002, according to Marsh. He also noted that Mary Jo Gibson is working with her son on an Eagle Scout project replacing Biblical plants in the church garden at St. James Lutheran Church near Wapwallopen in Pennsylvania and that Mary Lou Froehl of Sts. Peter and Paul Church in Petersburg, Indiana, also is planning a new garden there. In addition, Marsh has posted other good news that the Parkside Lutheran Church in Buffalo, New York, is hoping to install a Biblical garden with the encouragement of one of their deacons. In Greenville, Texas, master gardener "Pud" Kearns has designed a breathtaking Biblical garden right in the center of the city. You can see that dramatic design and get some ideas from it by visiting the Fair Haven website where Reverend Hudson-Knapp has posted this garden plan in graphic color. It is my hope that this book, and the inspired gardens you'll read about in it, will encourage many more people like yourself to dig in and begin growing Biblical gardens in the years ahead. Growing together with God is a worthwhile goal for every gardener.

Chapter Nine

World's Most Magnificent Biblical Garden

Neot Kedumim is a 625-acre Biblical landscape reserve in the Holy Land and the most remarkable and magnificent Biblical plant garden in the entire world. It is more extensive than any other Biblical garden and has a wider range of authentic Biblical plants and landmarks than anywhere else. Neot Kedumim presents a restored natural environment and plant habitat as gardens once were in Biblical times in the Holy Land. There are spiritual and gardening lessons there that are eminently transplantable to other gardens, even small backyard plots, all around the world.

Neot Kedumim began as a dream shared by Dr. Ephraim and Hannah Hareuveni, two Russian Jewish emigrants to Israel who envisioned these gardens of flora and fauna. Both were trained botanists and teachers who dedicated themselves to research of the land and ancient literature of Israel.

They conceived the idea of developing a living replica to reflect the interrelated botany, history, and traditions of the land of the Bible, which would create a bond between the past, present, and future. The Biblical Landscape Reserve in Israel thrives today on 625 formerly barren acres in Israel's Modi'in region as a pastoral network of Biblical and Talmudic landscape that attracts more than 100,000 visitors a year.

As Helen Frenkley, the director of Neot Kedumim astutely noted, it conveys messages through accounts of people interacting with a particular land. The word *neot* means "pastures" or "places of beauty," as in Psalm 23: "He maketh me to lie down in green pastures." The word *kedumim* means "ancient" and contains the Hebrew roots of this word that indicate forward movement in time that also expresses hope of future growth from past roots. That is a most apt and expressive name for this remarkable Biblical landscape reserve.

Growing there are the seven varieties of food that Moses mentions in Deuteronomy 8:8 as the most important crops of ancient Israel: wheat, barley, figs, pomegranates, grapes, olives, and dates. All plants along walkways and trails are marked with appropriate quotes from the Scriptures. In addition to the Biblical plants visitors see gleaming man-made lakes, 2,200-year-old cisterns, and Byzantine chapels. Ancient Roman period olive oil presses and other relics of 2 millennia and more ago are on display.

Visitors can rest under an elegant, historic Cedar of Lebanon tree. Even better, they can inhale the scents from the flowers, the herbs, the essence of earliest times. Special trails provide ideas for all who wish to explore the outreach potential of Biblical gardens at their churches or temples or organizations. The end of this chapter provides website information and how to enjoy a virtual trip there via the Internet.

Along one trail fields are brimming with flowers in season, and along the walk are several varieties of iris and an array of cyclamens, daisies, and Sharon tulips. Anemones, a classic flower of the Scriptures and Holy Land, provide their red, purple, and sometimes white delights. Still another trail guides visitors to vistas of plants where white lilies reflect the passage "his lips are like lilies, dripping flowing myrrh" from Song of Solomon 5:13. Along the way are buttercups, from Song of Solomon 2:12: "The red buttercups appear on the earth." Further along, people see grapevines and striking clusters of white and yellow narcissus, which reflect that Biblical image, "I am the narcissus of the valleys," from Song of Solomon 2:1.

In the fall, there is the "ingathering," a harvesting time for the last crops of the year and preparation for the coming planting season when fields are plowed and sown with seed grain in anticipation of the coming winter rains. White squill, meadow saffron, which is a type of crocus, and other flowers, herbs, and trees are part of all the walking trails. Along these trails, printed guides in various languages enable visitors to further enjoy and understand the important Scriptural, historic, and ecological significance of what they see growing. For example, you can find many of the key Biblical herbs there. (See table 9.1.)

Table 9.1 Herb Guide at Neot Kedumim

	Identification	
English Common Name	**Latin Name**	Biblical (B) / Talmudic (T)
balm of Gilead	(possible) *Balanites aegyptiaca*	tzori (B)
caper	*Capparis spinosa*	tzalaf (T)
carob, also St. John's bread	*Ceratonia siliqua*	kharuv (B,T)
hyssop	*Majorana syriaca*	ezov (B)
hyssop, tea	*Micromeria fruticosa*	ezov beit diklai (T)
lavender, Spanish	*Lavandula stoechas*	ezov kokhli (T)
myrrh (possible)	*Commiphora abyssinica*	mor (B)
oleander	*Nerium oleander*	harduf (T)
sage, Jerusalem,	*Salvia hierosolymitana*	moriah (B)
sage, Judean,	*Salvia judaica*	moriah (B)
sage, land of Israel,	*Salvia palaestina*	moriah (B)
sage, pungent,	*Salvia dominica*	moriah (B)
sage, 3-leafed,	*Salvia fruticosa*	moriah (B)
savory, whorled	*Satureja thymbra*	ezov romi (T)
thyme	*Coridothymus capitatus*	siyya (T)

Neot Kedumim continues to grow in many worthwhile directions today, thanks to the dedication and perseverance of Ephraim and Hannah's son, Nogah Hareuveni. He was the actual founder of Neot Kedumim, the Biblical Landscape Reserve, in

1965. Born in Jerusalem in 1924, from childhood Nogah worked with his parents, the founders of the Museum of Biblical and Talmudic Botany at the Hebrew University. For years he traveled afield throughout Israel with them as they collected and recorded the vast flora of the land of the Bible, so important to Judaism and Christianity.

Neot Kedumim remains dedicated to exploring and demonstrating the ties between the Biblical tradition and Israel's nature and agriculture as expressed in prayers, holidays, and symbols. Equally significant, it has deep roots in Judeo-Christian traditions. Years of devotion to this exceptional, living, and growing Biblical landscape project led Nogah to write several books: *Nature in Our Biblical Heritage* is the perfect gift for anyone interested in Bible, nature, history, or Judeo-Christian traditions. *Tree and Shrub in Our Biblical Heritage* and *Desert and Shepherd in Our Biblical Heritage* are equally worthwhile. All are available from American Friends of Neot Kedumim.

When Biblical garden enthusiasts travel to this reserve, in person or via the Internet website, they can wander among groves of palm, fig, and olive trees, observing the intertwining of Jewish and Christian traditions. Hundreds of transplanted ancient olive trees grace the reconstructed terraced slopes. In Exodus 27:20 God commanded that only pure olive oil be used to light the menorah, the 7-branched candelabrum that stood in the Holy Temple, because olive oil produces the brightest and steadiest flame of all oils. When the dove brought back an olive leaf in its beak to Noah on the ark, a homily relates that it was bringing the symbol of light and therefore of peace to the world. Two olive branches that surround the menorah are the emblem of the modern state of Israel.

Also prominent in the gardens is the moriah plant, which is a member of the sage family. This herb, which grows virtually wild throughout Israel today as it did in Biblical times, has special significance to the Jewish people. The menorah is first mentioned in the Bible when God instructed Moses in the preparation of the Ark of the Covenant. As described in Exodus 25:31–40, the specifications seem almost couched in botanical terms of branches, calyxes, cups, and petals. Ephraim and Hannah Hareuveni were the first to point to a direct relationship between the menorah and the moriah or sage plant as a particular Biblical plant. The moriah may not always have 7 branches, but it does have an even number growing from a central branch, and its pattern is strikingly similar to the menorah. See chapter 2 for more about sage.

Neot Kedumim is located in an area known in the Bible as the Judean foothills or lowlands. Centuries of overgrazing, battles, and neglect had eroded the land down to bedrock. To create gardens, tens of thousands of tons of soil had to be trucked in to the site and thousands of trees and bushes transplanted to their new/ancient home. This 625-acre reserve re-creates a series of landscapes and explores the intertwining between the land of the Bible and the texts of the Hebrew Bible and the Christian New Testament.

More than a decade ago, Neot Kedumim became part of Israel's educational system. Students from kindergarten through high school and all parts of the country's religious and social spectrum have been visiting Neot Kedumim annually during the past decade. Teachers from all parts of the country, religious and secular, young and old, regularly take part in educational programs. The programs have at times transcended religious differences as well. Four Moslem Arab teachers from a village in the Galilee participated in one of the courses. The following year they brought some of their colleagues to the program. Growing together in peace is a worthwhile theme at Neot Kedumim, in the Holy Land, and around the world and is the underlying theme in this book, too.

As you plan your Biblical garden, no doubt you will be concerned with both funding it well and developing worthwhile outreach programs. The folks at Neot Kedumim do both by serving meals created with foods that grew in the Holy Land in Biblical times. The meals allow visitors to "taste and see" Biblical foods. The proceeds from meals also help to fund operations much like church dinners help support religious programs, groups, and gardens in the United States.

Church suppers have been and are a traditional fund-raising activity for thousands of churches across America. Perhaps some of the special food and feast projects at this innovative Biblical garden will provide food for thought as fund-raiser projects for churches and temples here in the United States. After all, foods that trace their history to Biblical plants can have special meaning for special events. Neot Kedumim folks know that visitors get hungry. They have come up with tasty meals to feed the multitudes, or a few visitors at a time, as the case may be.

From Proverbs 15:17 we read, "Better a meal of vegetables where there is love than a fatted ox where there is hatred." In keeping with that passage Neot Kedumim offers vegetarian menus. You may not wish to become that deeply involved as operating a regular food service at your church or temple, but there may be some opportunities for special events, fund-raisers, or Holy Day suppers that come to mind as you read their menus.

Innovative ideas deserve mention and following is a typical menu from Neot Kedumim. Before dinner a visitor can savor appetizers that include garbanzo beans and olives in hyssop sauce. Cucumber and carrot strips dipped in tehina. Roasted pita bread with olive oil and garlic. In these meals herbs are one of the key ingredients, of course.

Then, visitors may choose from the buffet among flavored goat and sheep cheeses: natural, creamed, with olives, walnuts, dill, or onion, plus four kinds of fresh bread, reflecting the grains grown there in ancient times. Also, flavored butters: natural, dill, onion, or garlic, and pickled or marinated.

Salads include pickled quail eggs with herbs; yogurt and cucumber strips with dill and mint; tabouleh, which is cracked wheat with herbs; vegetables; and olive oil. Other choices are garbanzo beans with capers; chopped olives with parsley, garlic, lemon squares, and olive oil; spinach leaves with red lentils, garlic, olive oil, and mint; and sliced cucumbers with pickles, dill, and purple onion.

Hot dishes include squash and onion casserole with rice and lentils and thick lentil stew. Squash stuffed with rice and raisins and squash stuffed with cheese also are available. Among other supper treats are herb salad with parsley, coriander, mint, dill, and pine nuts; strips of squash, lightly cooked with hot and sour sauce; squash with onions and herbs; white beans with chopped spinach and onion rings in olive oil and lemon; lettuce salad, cabbage and sesame in sweet and sour sauce; raw carrots with walnuts and raisins; beets with walnuts and herbs; and grape leaves stuffed with rice.

Desserts include in-season fruit and dough puffs with choice of homemade jams and honey. Beverages are cold lemon and almond drinks and hot herbal tea. Wine is optional.

For those who wish a morning treat, a Biblical-style breakfast is offered as a buffet. It includes shepherd cheese, farmer's cheese, garden cucumbers, olives and olive oil, hyssop seasoning, fresh-baked pita bread, along with yogurt, date honey, bee honey, raisins, walnuts, and seasonal herbal teas. As noted in their literature, the menu seems appropriate to quote from Psalm 78:29, "and they ate and were well filled."

This delightful array of menu items offers lots of ideas for Biblical meals for fund-raising events at churches and temples with Biblical gardens in North America. In fact, their creative menus invoke ideas for special Biblical meals, whether your church has a Biblical garden or not. For example, the First Congregational Church in Fair Haven, Vermont, includes one or more Biblical food items in a more traditional church supper as a special highlight.

Another idea has taken root at Neot Kedumim and become increasingly popular. The Wedding Trail is a gentle walk through landscapes evoked in the Song of Songs. Proceeding from a pool reminding participants that "torrents of water cannot quench love, nor rivers drown it," the trail skirts young pomegranate trees, reflecting another Scriptural passage, "Your limbs are an orchard of pomegranates, with all choicest fruits."

After walking through the landscapes of the "tulip of Sharon," the "apple among the trees of the forest," and the "narcissus among the thorns," wedding guests arrive under the tall date palms beside the Pool of Solomon. The trail is illuminated at night and provides a truly romantic setting for a wedding. After the ceremony, the Neot Kedumim tour trail takes wedding guests to a reception area, a spotlit clearing in a pine forest for eating, singing, and dancing. That too has its Scriptural reflection from Isaiah 55:12, "For you shall go out with joy, and be led forth with peace; the mountains and the hills shall break forth before you into singing, and all the trees of the field shall clap their hands."

Your Biblical garden can serve as a site for weddings or other special events. It might also serve as a special site for wedding photos, with selected flowers and plants that may have a special significance to the event or people involved.

Another aspect of Neot Kedumim deserves special focus, again for ideas that may be useful as we see how others have overcome obstacles, creatively planted, and cultivated their gardens. According to Shlomo Teitlebaum, Neot Kedumim deputy director for development, the planners there faced some dilemmas while reconstructing the landscapes of the Song of Songs and had to come up with creative solutions.

"It was King Solomon's words in Ecclesiastes that gave us the inspiration and courage to develop a portion of the bare, rocky tract of the Dale of the Song of Songs. Two hills hugging a valley of varying width would be ideal for verses from the Song: 'His aspect is like Lebanon, noble as the cedars,' and 'I went down to the walnut garden to see the budding inside the stream banks,' and 'you stand stately like a date palm.'

"But how could we plant cedars and walnut trees which grow in high cold habitats, next to date palms and the narcissus of the valleys which need warmth and large quantities of water?" Teitlebaum asked rhetorically. "How could we put 'mountains of spices' next to the Sharon tulip, grazing areas for animals next to cultivated gardens of grapevines and pomegranates?

"Solutions to some of the problems were suggested by the words of the Song itself: 'I went down to the walnut garden.' Walnut trees normally grow at high elevations, but as the Hebrew verb 'went down' suggests, walnuts also can grow in a micro-climate created by a topographical depression surrounded by hills. The cold air sinking into such a pocket provides nights of winter frost which the walnuts need," Teitelbaum explains.

"Red sandy loam was brought from the Sharon plain. The topsoil, with the seeds of windflowers in it, was taken from an empty field just before apartment towers were built there. Along with the white broom and oaks of the Sharon habitat, we planted bulbs of the Sharon tulip.

"A pond was dug at the edge of the valley to catch runoff rainwater and provide a source for irrigating the walnut trees. Today, that vision has come to fruition and visitors can enjoy delicious ripe walnuts in the fall. Next to the pool, a large catchment basin was dug. Muddy soil that forms there during the rainy season provides a good habitat for 'narcissus among the thorns,' " Teitlebaum points out.

The pool, called appropriately the "Pool of Solomon," reflects the tall date palms planted beside it, as well as the cedars growing on top of the adjacent hill; the majestic trees that appear together in the lovers' words in the Song and in Psalm 92. " 'The righteous will flourish like the date palm, and grow tall like the cedar of Lebanon. And quite naturally, this is the setting for wedding ceremonies at Neot Kedumim,' " Teitlebaum concludes.

Neot Kedumim is located on Route 443 near Modi'in and the Ben Shemen Forest, only 10 minutes by car from Ben Gurion airport and 30 to 45 minutes by car from Jerusalem or Tel Aviv. Visiting hours are Sunday to Thursday, 8:30 A.M. to sunset, Friday, and Jewish holiday eves 8:30 A.M. to 1:00 P.M. It is closed Saturdays and Jewish holidays. Their colorful website, one of my favorites Biblical garden sites, is www.neot-kedumim.org.il. The website provides an enjoyable virtual walk into Biblical natural history. Click in and enjoy some tours today. You'll be as I am, captivated by the beauty of the Holy Land.

The best way to get information in North America is from Innes Kasanof, executive director of American Friends of Neot Kedumim, 813 Route 3, Halcott Center, NY 12430. The telephone is (845) 254-5031, fax is (845) 254-9836, and e-mail is afnk@catskill.net. All Neot Kedumim publications are available by direct order from AFNK. I trust that this book will also help encourage support for that marvelous project. As the olive branch stands for peace, may Neot Kedumim continue to grow and encourage more Biblical gardens to sprout and thrive in peace around the world in keeping also with the theme of this book, "Let's Grow Together."

OTHER BIBLICAL GARDENS AROUND THE WORLD

World of the Bible Gardens

In Jerusalem there is another garden, but unlike any you may have visited. This World of the Bible Gardens has full-scale archaeological replicas that help interpret the Scriptures. There is a goathair tent, a real sheepfold, a stone manger, and a well from which visitors can draw water.

There also is a watchtower, wine press, a threshing floor, and olive press. At the stone quarry, visitors can learn about ancient building methods. The Scripture Garden is located in Ein Karem, Jerusalem, Israel, and offers classes and even has authentic Biblical meals. With all these attractions, the plus is the plants that are growing at this unusual site.

The Garden of Gethsemene

Another special garden in the Holy Land is worth noting. It is located at the western base of the Mount of Olives. Within the walled perimeter of the Church of All Nations is a small grove of ancient olive trees. The historic Garden of Gethsemene holds a place in the hearts of all Christians as the site where Jesus prayed the night before he was taken captive as described in Mark 14:32–50. In this churchyard, local tradition dates these ancient trees back to Biblical times, but that is unlikely. More probable is that the gnarled and twisted olive trees may be descendants of the original ones. They are undoubtedly very old. Although not a typical Biblical garden this unique spot deserves its place in this book as a special garden landmark in the Holy Land.

For those who wish to wander around the world to visit other fine Biblical gardens, here are some that I've located during my research for this book. A few had websites at the time of publication and I've included them in my favorite Biblical plant list (see chapter 11) at the end of this book.

More Biblical gardens seem to be taking "root" every year. In the meantime you may try some explorations via www.google.com to see what you can find. More gardens are being added periodically. Happy Biblical garden touring around the world.

- Bible Garden Memorial Trust, Palm Beach, New South Wales, Australia
- Biblical Garden, Rockhampton Botanic Garden, Queensland, Australia
- Biblical Garden, St. Paul's Cathedral, Sydney, NSW, Australia
- Biblical Garden—Brickman's Country Gardens, Ontario, Canada
- Biblical Garden, Holy Trinity Parish Church, Sheen Park, Richmond, Surrey, England
- Millennium Bible Garden, St. Mark's, Isle of Man, England
- Millennium Bible Garden, Woodbridge, Suffolk, England
- Biblical Garden, Royal Botanic Garden, Kew Gardens, Richmond, Surrey, England
- Biblical Garden, Rutland Street Chapel, Christchurch, New Zealand
- Biblical Garden, Sternberg Centre for Judaism, The Manor House, London, England
- Biblical Garden, Amsterdam Free University, Amsterdam, Holland
- Biblical Garden, St. Benedict's Priory, The Mount, Cobb, County Cork, Ireland
- Biblical Garden, St. George's College, Jerusalem, Israel
- Yad-Hasmona Biblical Gardens, Israel

Franciscan Center of Environmental Studies, Rome, Italy
Biblical Garden—Elgin, Scotland
Biblical Garden, Bangor Cathedral Close, Bangor,
  Caenaarvon, Wales

Chapter Ten

Biblical Plant Celebration Service

Genesis 1:11–12 says, "In the beginning, God created the earth and the waters and then He created the plants on the third day"; the beginnings of faith and our first discoveries about plants are found here.

"Let the earth bring forth grass, the herb yielding seed, and the fruit tree yielding fruit after his kind, whose seed is in itself, upon the earth: and it was so. And the earth brought forth grass, and the herb yielding seed after his kind, and the tree yielding fruit, whose seed was in itself, after his kind: and God saw that it was good."

As you read these marvelous words in Genesis, describing the creation, you find the earliest references in the Scriptures to plants of the Bible. As I wrote my first book about Biblical plants, *Your Biblical Garden,* in 1980, little did I realize where that would lead. It has taken me on a long journey of research into the Bible, about Biblical plants and their Scriptural references, and also into history of the many different plants of the Holy Land.

It seemed appropriate to use what I have learned and been taught by many Biblical scholars for a meaningful church service that could be given by others at their respective houses of worship. It is included in this chapter. This project began as a

Biblical Plant Celebration Service in collaboration with my longtime friend, Reverend E. Lamar Robinson, minister of Christ Church in Kennebunk, Maine. We wanted to celebrate God's gifts from the earth and the miracle of plant life, from seed or bulb or root to growth and bloom and beautiful, bountiful harvest.

We also wanted to help make the Bible come alive by showing that we share some of the same plants as our spiritual forefathers. And, we wanted to help reduce the distance between Bible lands in history and the Holy Land today, as we reveal that we all can grow together with the plants of the Scriptures in backyard, churchyard, school yard, and community. As we embark on new growing experiences with plants of the Bible, it is well to remember that what we have is truly a gift from God. As you sow your herb seeds and flower bulbs, plant your fruits and trees; you are part of the continuing creation of life and beauty. We are all, in fact, gardening with God.

Today, with this simple, down-to-earth Biblical Plant Celebration Service, you can celebrate the blooming beauty of Biblical flowers, herbs, vegetables, fruits, wherever you live. I hope you will include others in this worthwhile project: your extended family of friends, neighbors, and members of your church and community. Perhaps you too may join with others to actually plant a Biblical garden as suggested in an earlier chapter. We all can dig in together to grow and enjoy the fruits and vegetables, flowers, herbs, and trees not only for their beauty, fragrance, nutrition, and taste but also for the pleasure that they give you from their special meaning as plants of the Bible, rooted in the Holy Land and Scriptures.

Following is a basic outline for needs. You and your religious leaders can adapt this as you wish. It includes suggested hymns and readings for lay members of the congregation and others. Please feel free to use this as your guide to a meaningful Biblical plant service.

A CELEBRATION SERVICE OF PLANTS OF THE BIBLE

Silent Prayer
Prelude—With Verdure Clad—Haydn
Call to Worship
Opening

Today, in this Special Celebration of Plants of the Bible, we wish to celebrate God's gifts from the earth; to behold with wonder the miracle of seeding time and harvest. Today, with the same reverent attitude that characterized our spiritual ancestors of Biblical times, we marvel at the orderly process through which God brings forth from the earth both food for our nourishment and beauty for our enjoyment. Today, as we realize that many of the plants that we can grow in our yards have ancestors rooted in our Biblical heritage, we may closely identify with those people through whom God spoke his eternal message to all succeeding generations.

Leader

The earth is the Lord's in all its beauty, the world and all who make their home upon it.

People

For God has made the earth, with its fertile valleys and mountain peaks, its green forests, fields, plains; its wide seas and great rivers, its infinite variety of life.

Leader

Who shall take his rightful place among all that God has created? To whom shall God say, "You are a faithful keeper of my earth?"

People

Those who appreciate the world God has given us. Those who use, but do not waste its resources, who enjoy but do not spoil its beauty, who are wise and thoughtful tenants.

Leader

Blessed are those who know the earth is not theirs, but belongs to God, who has given it for the benefit of all.

People

Let us praise God and be joyful; let us rejoice in the green earth, the warm sun, winter's fierce breath and the gentle touch of spring; all living, growing things. Let us rejoice and be glad for the earth is the Lord's in all its beauty, the world and those that make their home upon it.

HYMN "Morning Has Broken"

Leader

Lord, today our thoughts go far beyond these four walls. We think of your garden we call the earth and its orderly process of seed and harvest; the cycles of night and day, sun and rain; winter, spring, summer, fall. We rejoice in the blooming beauty of flowers, the productiveness of vegetable gardens, the spicy smell of herbs, the gracefulness of trees. We think of the good earth you have given us to enjoy, use, protect, and preserve. When we reflect upon these things, we are grateful, we are joyful, we are glad, and with the Psalmist we say, "The world is filled with the glory of God."

HYMN "For the Beauty of the Earth"

Reflections about Biblical Flowers

Leader

"Hyacinths and biscuits," wrote the poet. Beauty and utility. Who is to say we need hyacinths less than biscuits, if our soul is to soar and our spirit to flourish? Beauty is not God, but surely one of the attributes of God is beauty. God gives us beauty in many ways, the glorious sea in its splendor, rugged mountain peaks, white rushing streams, gleaming church steeples pointing to the heavens, the glory of a simple flower in your backyard, lifting up its face to the sun and a fragrant herb that scents the air.

"What good is it?" the strict pragmatist may ask gruffly, looking at a lily. The whispered answer: "To show forth the glory of God."

"Oh," he slowly replies, seeing for the first time ever a new dimension in God's world and in his own life.

A free-verse poem by Alfred Lord Tennyson says what a lot of us think, not only about flowers, but all growing plants and trees. He wrote:

> Flower in crained wall,
> I pluck you out of the crannies.
> I hold you here, root and all, in my hand.
> Little flower, but if I could understand
> What you are, root and all, and all in all,
> I should know what God and man is.

Gardener

The Bible is alive with mentions of flowers as they grew in the Holy Land. [At this point, you can add several Scriptural References about Biblical plants There are many included in this book, so you can choose from them or add others as you wish.]

You can grow many of these glorious flowers right in your own yard. Because of the two-season climate of the Holy Land, wet and dry season, most of the flowers grow from bulbs. You probably already know many of them. Crocus, anemone, hyacinth, narcissus, lily, iris, and others. [You may add Scriptural references as you wish from my *Flowers of the Bible* book.]

Reflections about Biblical Vegetables

Leader

During the past decade, millions more American families have dug in and begun enjoying the rewards of growing vegetables. Many of you probably are vegetable gardeners, even if it is only a row of lettuce, a hill of cucumbers or squash, or a few tomato plants. There is something almost spiritual about getting your hands down in the good earth—planting, tending, and cultivating your crops. What can compare with the satisfaction of harvests? You have been a coworker with God. Your dinnertime grace will have a new meaning when the vegetables have come from your own work in your own garden. You may even wish to offer the words of one of the most ancient of Hebrew blessings; a part of the ritual of Passover: "Blessed are you, O Lord God, who brings forth food from the earth."

Gardener

In Genesis, we read about the creation. And later, as we read our Bibles, we find references to the vegetables, cucumbers, melons, leeks, onions, and other foods. All of these plants are deeply rooted in the Bible and Holy Land. Happily, we can trace their roots and plant them, grow them and harvest tasty cucumbers, melons, and other vegetables from our own gardens.

Reflections about Biblical Herbs

Leader

What is an herb, but a plant with a flavor and a smell all its own, which, when used in the right proportions, improves in a delightful way the food with which it is cooked! It is as if God wished to let us "spice up" our life a little and gave us herbs to use in order to add flavor and variety to our cooking. A woman who teaches at a famous hospital had a story about herbs that bears retelling. She had taken several mentally ill patients to a garden where herbs were growing. Some of them had retreated so far from the world that they didn't even talk, but who began to speak again after smelling certain herbs with which they had good associations in their earlier years. Perhaps the aroma of the herbs reminded them of their mother's kitchen in their early childhood, in the age of innocence, before the complexity of the world overcame them.

The patients would sometimes ask whether they could bring a leaf back with them to their rooms. It was, she recalled, the first, small but crucial step on the road to their recovery. There is power and mystery in herbs, now as in the days of our spiritual Holy Land ancestors. Herbs are one of God's gifts from the good earth.

Gardener

We find the first mention of herbs in Genesis [add full or partial quotation as you wish from the Bible] and often again through many passages and pages in the Bible. Several of them are easily and tastefully grown in our own gardens: coriander, dill, hyssop, mint, rosemary, sage, and many more. Herbs are making a deserved comeback in American gardens. Those that have their roots in the Scriptures and Holy Land impart their special meanings for us all.

[Here you may wish to refer to some specific herbs, using the Scriptural quotations about them that you find in this book, and even provide some tips for growing them to encourage parishioners to dig in to the good earth and grow some Biblical plants themselves.]

Reflections about Biblical Fruits

Leader

We have all read and know about the fruits of the Holy Land—dates, figs, olives, grapes. We may not know how they grow or be able to grow many of them ourselves, but we can enjoy their good flavor. Life is like that. Everyone doesn't have to know it all. Everything doesn't have to grow everywhere. There is an interdependence in life that is wholesome, and we can see and appreciate it even in something as basic as the geographical distribution of fruit trees and vines.

In Biblical literature, vineyards and olive groves are symbols of the well-being of the people. They are often used to express, symbolically, the bountifulness of God's favor, and the graciousness of God's gifts. Let us reflect that while dates and olives are right for growing in other lands and places, we can see and enjoy the fruitful abundance of that other popular plant of the Holy Land, grapes, hanging in abundance from our own backyard arbor, as we reflect on this fruitful gift that God has given us.

Gardener

Grapes are one of the most often mentioned plants in the Bible, if you include the many references to wine, which was one of the common beverages of that era. Today, everyone can enjoy many types of grapes—green, red, yellow—that offer tasteful bounty from arbors in our own backyards.

Reflections about Biblical Trees

Leader

There is a special quality about trees which is sometimes difficult to put into words. If you have stood beside a giant California redwood, remember that it was a sapling when a man named Jesus walked with his disciples along the Sea of Galilee. That same tree was quite mature when a man named Columbus was believed to be crazy for thinking that the world was round. The tenacity of trees just below timberline in the mountains is really remarkable; they hang on doggedly against the odds, fighting the cold of winter and the dry heat of summer—conditions of severe wind and space and rocky soil. Or, think of cypress trees with their bases hidden beneath the dark waters, their knees protruding up as if for a breath of air.

We can all enjoy the sights of trees and the scents of them, too, the distinctive aroma of olive wood from the Holy Land or the heady scent of cedar. Who doesn't appreciate the strong, sweet smell of balsam firs during the holidays or the pungent smells of a pine grove after a rain?

Trees are a glorious expression of God's graciousness as creator. As we look at the trees in our yards, around our town, and in the woods where we camp on vacations, we may see them as messengers of God's greatness and goodness.

Gardener

Many different trees are mentioned in many passages in the Bible; for example, almond, cedar, laurel, pine, poplar, oak, willow, and walnut. You can enjoy the sight of some of them that you can plant and grow in your own landscape that are related to the trees of the Holy Land. In fact, growing trees is one way we can all beautify our home landscapes, improve our shared environment, and also add to the enjoyment of generations to come.

Leader

Let us reflect on the flowers, herbs, vegetables, fruits, and trees that we can plant and enjoy, as we refocus on learning to grow together better, here in God's world today, and for the years ahead. We are all indeed stewards of this land called earth and are truly "Growing with God" as we should be.

HYMN "This Is My Father's World"

Closing: Thoughts, Reflections, Benediction.

Chapter Eleven

Favorite Biblical Garden Websites and Plant Sources

I wrote *Herbs of the Bible* after a book on Biblical flowers. When I was finished, I realized several more Biblical gardens had sprouted. I also discovered that more of these gardens are posting websites with glorious photos and more useful information about the plants, including herbs, than ever before. In the process I've been taking courses in computers and getting help from our friendly local computer specialists. Librarians are another good source for learning computers and the Internet. There is a wide world to discover on many topics on the expanding Internet and, if you have not already done so, I strongly recommend that you learn more about it. Most important, gardening information is being made available that makes growing plants of all types easier and more fun. Visit with expert gardeners, ask questions, and see the gardens in all their glory.

As more Biblical gardens sprout, bloom, and grow across America and around the world, their founders and tenders want to share what they grow with you and millions of other potential Biblical gardeners and people of faith. Thanks to the Internet and cyberspace, today you can visit some of them with a click of your mouse. Many have glorious photos of flowers and landscapes plus trees, herbs, fruits, and vegetables. Even better, there is a wealth of valuable information about the plants they grow, links to other Biblical gardens, and plant information and beautiful color photos to enjoy too. Each reflects the dedication and love that the founders, directors, webmasters, and those who tend them faithfully have invested and willingly share.

Sometimes useful new research and gardening tools seem to appear overnight, contributors add new, wonderful, and worthwhile information. One of the best and most thorough websites is www.biblicalgardens.com. There you'll find an abundance of knowledge, sources for plants, seeds, and a chance to visit with other Biblical gardeners, both novices and veterans, with great ideas and wide experience to share.

Biblical garden friends sent a relatively new one along so I have added it here as we go to press. The goal of www.gardenprojectonline.com is to activate peoples' faith using the garden as a "teaching tool," according to the founder and faithful gardener Steve Hallowell. He started the project because he felt that more people could be attracted to a relationship with Christ if they could relate it to something familiar like gardening. As one visitor put it, "you are really trying to give faith a little visual substance to go on." Much like Jesus used parables, www.garden Project Online is trying to activate peoples' faith. As Steve Hallowell says, "My goal is to get people up 'out of the pews' and into an everyday relationship with God. Gardening is a wonderful vehicle to soften peoples' soil, and is a living symbol of God's miracles." To that thought and many he expresses so well I add, "God bless you, Steve," and may you continue to provide inspiration, motivation, and gardening help to all.

As Shirley Sidell has done so well with her website, it is important that we people, of every faith and denomination, learn to share in our gardens and in our lives. As we do we can help others approach a better garden and life, and with hope, prayer, and work that may help us all on the path to peace.

For decades, and in this book, my living and writing theme has been "Growing with God," helping to share Biblical gardening ideas, tips, and advice gathered from many gardeners with readers everywhere. I always welcome information about other Biblical gardens and gardening ideas that deserve to be celebrated and saluted. Please feel free to send me e-mail about such gardens with the names of people who tend them at aswenson@gwi.net so that they may receive the recognition they deserve when I revise the next edition of this book.

Following are those marvelous Biblical garden websites that you can use to expand your Biblical gardening horizons. With my thanks to all who have helped me with this book and are so willing to share great growing ideas, advice, plus their wit and wisdom.

MY FAVORITE BIBLICAL GARDEN WEBSITES

Bible Garden Links for Adults
www.suite101.com/articles

Biblical Garden
Elgin, Scotland
www.dufus.com/Duffus+2000biblical_garden/

Biblical Garden References
Shirley Sidell, Founder
www.biblicalgardens.com

Brinkman Country Gardens, Canada
www.granite.sentex.net/~lwr/brickman.html

Cathedral of St. John the Divine
New York, New York
www.stjohndivine.org

First Biblical Resources U.S.A.
www.brusa.org/biblegardenpage.htm

First Congregational Church
Fair Haven, Vermont
www.sover.net/~hkfamily

First Presbyterian Church
Gainesville, Florida
www.gnv.fdt.net/~hpc/garden/

First Presbyterian Church
Nashville, Tennessee
www.klis.com/scove/041400.htm

Magnolia Plantation and Gardens
Charleston, South Carolina
www.magnoliaplantation.com/gardens/index.html

Neot Kedumim
Lod, Israel
www.neot-kedumim.org.il

Paradise Valley United Methodist Church
Paradise Valley, Arizona
www.pvumc.org/about/biblicalgarden.html

Rodef Shalom Biblical Gardens
Pittsburgh, Pennsylvania
www.rodefshalom.org/Garden/initial.htm

St. John's Episcopal Church
Norman, Oklahoma
www.episcopalnorman.org/outreach.htm#garden

Temple Beth Shalom
Sun City, Arizona
www.goodnet.com/~tbsaz/body_index.html

Temple Sinai Biblical Gardens
Newport News, Virginia
www.ujcvp.org/temple_sinai/bulletin04-00

Warsaw Biblical Gardens
Warsaw, Indiana
www.warsawbiblicalgardens.org

Yad-Hasmona Biblical Gardens, Israel
www.yad8.com/bg/biblical.htm

Finally, a most helpful website for the study of the Bible, as well as comparisons and research of many topics and Scriptural references in a variety of versions of the Bible, can be found at: www.biblestudytools.net.

FAVORITE BIBLICAL PLANT SOURCES

Finding appropriate Biblical flowers has often been a difficult task for those starting out with a plan, goal, and vision, but only a few bulb, seed, and plant sources. After talking with many accomplished Biblical gardeners and scanning more than 60 mail-order catalogs, I have compiled a list of sources that have proved reliable and reasonable, and also have a range of varieties from which to select those that are best suited for Biblical gardens. Also, because these established mail-order companies are growing plants themselves, it is worth checking their websites periodically for new flowers and also useful growing tips. Naturally, you may find many Biblical herbs and plants locally, especially among the thousands of herbalists who exist all around the country. Because of the new popularity in herbs, many national chain stores now offer quality herbs as prestarted plants each spring. I found several more mint and basil varieties in the expanded garden department of a national chain this year to add to my collection. Then, I discovered a hobby herb garden only a few miles from my home. Sometimes it pays to keep one's eyes wide open and drive a few back roads nearby. In the meantime, here are some of the more reliable mail-order companies featuring Biblical plants including herbs. Some obviously feature only bulbs and fruits. They are included in this book about Biblical herbs because you may begin with a few herbs or flowers, but as Biblical gardening becomes more popular, this wider list may serve as a good reference for your growing horizons.

Bluestone Perennials
7211 Middle Ridge Road
Madison, OH 44057
(800) 852-5243 or (440) 428-7535
www.bluestoneperennials.com

Reasonably priced and good selection. Sarah Boonstra is very helpful.

Dutch Gardens
P. O. Box 2037
Lakewood, NJ 08701
(800) 818-3861
www.dutchgardens.com

A longtime supplier of quality flower bulbs, anemones, crocus, cyclamen, iris, lilies, tulips, and other fascinating flowering wonders. Joop Visser is the senior field manager.

Forest Farm Nursery
990 Tetherow Road
Williams, OR 97544

A new recommendation by several different Biblical gardeners. Write for their catalog.

Johnny's Selected Seeds
Foss Hill Road
RR 1, Box 2580
Albion, ME 04910
(207) 437-4301
www.johnnyseeds.com

Johnny's Selected Seeds is a reliable supplier of flowers, herbs and vegetables focused on organic production methods.

J.L. Hudson Seedsman
Star Route 2, Box 337
La Honda, CA 94020

Suggested by several Biblical gardeners. An ethnobotanical catalog is available.

Meadowsweet Herb Farm
North Shrewsbury, VT 05738
(802) 492-3565

Specializes in herbs, but some of those also bear lovely flowers, as sage does.

Mellinger's
2310 W. South Range Road
North Lima, OH 44452
(800) 321-7444
www.mellingers.com

Has a variety of flowers from seeds and bulbs.

Miller Nursery
Canandaigua, NY
(800) 836-9530

Reasonably priced, hardy fruit, nut, and foliage trees and shrubs. I've bought from them for years with good results.

Mountain Valley Growers
38325 Pepperweed Road
Squaw Valley, CA 93675
(559) 338-2775
www.mountainvalleygrowers.com/bibleherbgarden

Mountain Valley Growers has many different biblical varieties by mail order and features Biblical herbs and sets.

Old House Gardens
536 Third Street
Ann Arbor, MI 48103
(734) 995-1486
www.oldhousegardens.com

Heirloom bulbs with rare old hyacinths, narcissi, tulips, and others. Unique, informative catalog.

Park Seed
1 Parkton Avenue
Greenwood, SC 29647
www.parkseed.com

A large mail-order company with wide selection of flowers and seeds of less available varieties.

Quality Dutch Bulbs
13 McFadden Road
Easton, PA 18045

Many varieties and types of flower bulbs.

Richters
Goodwood, Ontario, Canada
www.richters.com

Recommended by several gardeners.

Select Seeds, Antique Flowers
180 Stickney Hill Road
Union, CT 06076

Delightful, hard-to-find flowers with informative catalog. Marilyn Barlow has assembled an excellent array of old-time heirloom flowers, Biblical ones, and many others worth growing for fun.

Seeds of Distinction
P.O. Box 86
Station A
Toronto, ON Canada M9C 4V2
(416) 255-3060
www.seedsofdistinction.com

Many hardy flowers and Biblical types.

Territorial Seed Company
P.O. Box 158
Cottage Grove, OR 97424
(541) 942-9547
www.territorial-seed.com

A treasure of information and good seeds at fair prices.

Van Bourgondien
245 Route 109
P. O. Box 1000
Babylon, NY 11702
(800) 622-9997
www.dutchbulbs.com

A major, longtime flower bulb company with many Biblical plants. Has useful information at website and offers planting tips, advice. Debbie Van Bourgondien is known nationally as the "Bulb Lady," a knowledgeable expert.

Wayside Gardens
1 Garden Lane
Hodges, SC 29695
(800) 845-1124
www.waysidegardens.com

Wayside Gardens has quality plants of harder to find species. Introducing exceptional Hellebores.

White Flower Farm
P. O. Box 50
Litchfield, CT 06759
(800) 503-9624
www.whiteflowerfarm.com

Specialists in rare bulb flowers, they recently purchased Daffodil Mart with its many rare, heirloom-type flowers.

Chapter Twelve

Greatest Global Gardening Sources and Favorite Garden Catalogs

To help you grow better with God, I've compiled my best gardening information sources for you in this book from all across America and around the world. Even better, you can tap top gardening talent at dozens of colorful, informative websites on the Internet and print out pages of valuable garden ideas, tips, and advice. If you haven't surfed the Web in search of great gardening ideas and advice, you're missing bushels of useful information and hundreds of free pages about every aspect and type of gardening.

You'll also find a wealth of herbal knowledge at herb companies, organizations, and other resources. Take a peek at beautiful flowers that you may wish to grow, learn about their habits and best tips for growing each.

The Mailorder Gardening Association is the world's largest group of companies that specialize in providing garden products via mail order and online. Many offer herbs. Some actually specialize in them. At MGA's periodically revised website, www.mailordergardening.com, you also can find a glossary of gardening words and phrases. They have more than 130 members who offer colorful, illustrated catalogs packed with tips and ideas.

The best thing about this site is the impressive list of member companies. They're categorized by the types of products they provide—annuals, perennials, herbs, fruit trees, garden supplies, fertilizer. Click on to a company name and up pops a short description of the catalog's offerings, phone number, and address. Click on the hotlink and you're instantly connected to the catalog's homepage. This website is a great place to start when you want to see just how many garden seed, plant, product, and accessory companies are online. You also can find toll-free numbers of companies that offer personal advice to your own gardening questions. That's especially useful when you are adding new herbs and plants to your garden and want facts about new varieties and practical how-to tips, too. Plus, you can then track down harvest, use, and even recipe sources. This wonderful world of adventure is right at your fingertips. And, libraries now offer courses and help so you can wander the world to find herbal information if you wish.

The National Gardening Association, www.garden.org, publisher of the *National Gardening* magazine, has one of the most comprehensive and useful sites on the Web. Hot buttons let you read articles from the magazine, find out about the NGA's Youth Garden Grants program, search the extensive NGA Library for data on a wide variety of gardening topics, and even check out a "seed swaps" section.

You also can subscribe to a free e-mail newsletter or ask gardening questions. Actually, 16,000 questions have already been answered, so there's a good chance that the information you want is already there waiting for you.

When you need just the right tool or piece of equipment and can't find it locally, check the Gardener's Supply Company site, www.gardeners.com. This is devoted primarily to fine merchandise found in their printed catalogs and periodically includes bargains not in the catalog. In addition, you can search an extensive Q&A library of gardening information.

Better yet, you can wander the Biblical plant world, too, and a 625-acre Biblical garden preserve in the Holy Land at www.neot-kedumim.org.il. Neot Kedumim is a labor of love, recreating an extensive preserve with the flowers, herbs, vegetables, fruits, trees, and also some of the fauna mentioned in Scriptures.

For organic gardening, which is immensely popular today, log on to the Organic Gardening website at www.organicgarden.com. They're America's experts and advice is plentiful.

As you surf the Web, you'll find hundreds of articles, info pages, and other material you may wish to save. Most sites allow you to download and print out these pages of helpful ideas, tips, and advice that are of interest to you. Be aware, of course, that sites and e-mails change as do people and contacts at these companies. Fortunately, new gardening companies are "sprouting and growing," especially those with specialty plants and products. A time-honored Biblical saying guides us to new growing ground: "Seek and Ye shall find," Luke 11:9.

MY FAVORITE GARDEN WEBSITES

Antique Flowers	www.selectseeds.com
Burpee	www.burpee.com
Cook's Garden	www.cooksgarden.com
Clyde Robin Wildflower Seeds	www.clyderobin.com
Drip Rite Irrigation	www.dripirr.com
Gardens Alive	www.GardensAlive.com
Gardener's Supply Company	www.gardeners.com
Garden to the Kitchen	www.gardentokitchen.com
Harris Seeds	www.harrisseeds.com
Johnny's Selected Seeds	www.johnnyseeds.com
Lilypons Water Gardens	www.lilypons.com
Mail Order Gardening Assn.	www.mailordergardening.com
Mountain Valley Growers, Inc.	www.mountainvalleygrowers.com
National Gardening Assn.	www.garden.org
Neot Kedumim, Israel	www.neot-kedumim.org.il
Nichols Garden Nursery	www.pacificharbor.com/nichols
Old House Gardens	www.oldhousegardens.com

Park Seeds   www.parkseed.com
Stokes Tropicals   www.stokestropicals.com
Thompson & Morgan   www.thompson-morgan.com
Van Bourgondien Bulbs   www.dutchbulbs.com
White Flower Farm   www.whiteflowerfarm.com
Wildseed Farms, Ltd.   www.wildseedfarms.com

MY FAVORITE GARDEN CATALOGS

Appalachian Gardens
P.O. Box 82
Waynesboro, PA 17268
(rare trees/shrubs)

The Cook's Garden
P.O. Box 535
Londonderry, VT 05148
(special salad/veggie varieties)

Bluestone Perennials
7211 Middle Ridge Road
Madison, OH 44057
(nice variety)

Crystal Palace Perennials
P.O. Box 154
St. John, IN 46373
(water garden plants)

Burgess Seed/Plant Co.
904 Four Seasons Road
Bloomington, IL 61701
(bulbs, seeds)

Drip Rite Irrigation
4235 Pacific Street, Suite H
Rocklin, CA 95747
(irrigation supplies)

Burpee
300 Park Avenue
Warmonster, PA 18974
(seeds/bulbs/plants)

Dutch Gardens
P.O. Box 200
Adelphia, NJ 07710
(Dutch bulbs)

Clyde Robin Seed Co.
P.O. Box 2366
Castro Valley, CA 94546
(wildflowers)

Ed Hume Seeds, Inc.
P.O. Box 1450
Kent, WA 98035
(short season varieties, plus)

Flowery Branch Seeds
Box 1330
Flowery Branch, GA 30542
(rare/heirloom/medicinal)

Forest Farm Nursery
990 Tetherow Road
Williams, OR 97544
(good source)

From Garden to Kitchen
1061 Elk Meadow Lane
Derry, ID 83823
(books/tools)

Gardener's Supply Co.
128 Intervale Road
Burlington, VT 05401
(many gardening supplies)

Gardens Alive
5100 Schenley Place
Lawrenceburg, IN 47025
(organic gardening source)

Harris Seeds
60 Saginaw Drive
Rochester, NY 14692
(old line seed firm)

Henry Field's
415 N. Burnett Street
Shenandoah, IA 51601
(plants/seeds/shrubs)

J.L. Hudson Seedsman
Star Rt. 2, Box 337
La Honda, CA 94020
(Biblical plants)

Johnny's Selected Seeds
RR 1, Box 2580
Albion, ME 04910
(seeds, wide variety list)

Klehm's Song Sparrow Perennial Farm
13101 East Rye Road
Avalon, WI 53505
(specialties)

Lilypons Water Gardens
P.O. Box 10
Buckeystown, MD 21717
(great water garden source)

Mantis
1028 Street Road
Southampton, PA 18966
(tillers/tools)

Mellinger's
2310 W. South Range Road
North Lima, OH 44452
(variety)

Miller Nurseries
5060 West Lake Road
Canandaigua, NY 14224
(great berry/fruit tree source)

Nichols Garden Nursery
1190 N. Pacific Highway NE
Albany, OR 97321
(many Asian, international)

Northwoods Nursery
27635 S. Oglesby Road
Canby, OR 97013
(rare fruits/nuts/others)

One Green World
28696 S. Cramer Road
Modalla, OR 97038
(rare international plants)

Park Seed
1 Parkton Avenue
Greenwood, SC 29647
(major U.S. seed and plant firm)

Quality Dutch Bulbs
13 McFadden Road
Easton, PA 18045
(many bulb flowers)

Roris Gardens
8195 Bradshaw Road
Sacramento, CA 95829
(iris specialists)

Royal River Roses
P.O. Box 370
Yarmouth, ME 04096
(rare, hardy, old-time roses)

Seeds of Distinction
P.O. Box 86
Station A
Toronto, ON Canada M9C 4V2
(unique)

Select Seeds Antique Flowers
180 Stickney Road
Union, CT 06076
(heirloom seeds/plants)

Stokes Seeds
Box 548
Buffalo, NY 15240
(many varieties)

Stokes Tropicals
P.O. Box 9868
New Iberia, LA 70562
(exotic tropical plants)

Van Bourgondien
P.O. Box 1000
Babylon, NY 11702
(major Dutch bulb specialist)

Vessey's Seeds, Ltd.
P.O. Box 9000
Calais, ME 04619
(U.S./Canadian varieties)

Wayside Gardens
Garden Lane
Hodges, SC 29695
(major plant source)

White Flower Farm
P.O. Box 50
Litchfield, CT 06759
(specialists in rare bulb flowers)

Wildseed Farms
525 Wildflower Hills
Fredericksburg, TX 78624
(wildflower specialists)

    Hundreds of free garden catalogs are currently available, and one handy garden catalog guide contains descriptions and contact information for 125 garden catalog companies and garden magazine publishers. Also included are smart shopper tips and a glossary of gardening terms. To receive a copy, send $2.00 check or money order to Mailorder Gardening Association, Dept. SC-AS, P.O. Box 2129, Columbia, MD 21045. This guide also lists dozens of garden companies with toll-free telephone numbers, advice hotlines, e-mail addresses, and websites to browse colorful pages at your leisure.

Glossary

Whether you are a veteran gardener or novice, plant and gardening terms can be confusing, especially when you are trying to describe some of your plants to another person. The following list provides plant terms, definitions, and a glossary of basic words. These are the terms I've used during the years in my syndicated garden columns and other books to make gardening simple and understandable.

acid soil Soil with a pH value less than 7 on the pH scale of 1 (acid) to 14 (alkaline).
alkaline soil Soil having a pH value greater than 7.0 on a soil test.
annual A plant that completes its life cycle from seed to mature plant to seed again in one growing season.
anther The polliniferous part of a stamen.
bacteria Microscopic, one-celled organisms lacking chlorophyll that multiply by fission and live on nonliving organic matter, helping to break it down into humus.
biennial A plant that requires two growing seasons to complete its life cycle and then dies. Vegetative growth takes place the first year and flowering and fruiting the second year.

blood meal An organic source of nitrogen that contains approximately 10 to12 percent nitrogen.

bud Naked or scale-covered embryonic tissue that will eventually develop into a vegetative shoot, stem, or flower.

bulb An underground stem that stores energy in modified leaves as found in bulbs of daffodils and tulips.

chemical fertilizer "Man-made" fertilizer made without carbon or derived from nonliving material, e.g., nitrogen from the air, phosphorous and potash from mineral deposits.

compost The end produce of aerobic decomposition of organic matter—plant residues, weeds, manures, lawn clippings, and other natural materials.

corm An underground stem that stores energy in modified stem tissue, e.g., a crocus corm.

cultivar Synonymous with variety, except that it refers only to cultivated plants.

decomposition Breakdown of organic materials into their constituent parts owing to the action of bacteria and microorganisms.

drip irrigation A system of adding water to a garden soil in a slow, gentle stream from sources such as a perforated hose or one with a bubbler or slow-release action. This watering method prevents soil from being disturbed and reduces water runoff.

foliar fertilization Application of plant nutrients directly to leaves rather than to soil. The leaves absorb nutrients to feed the plant.

forcing A process of making plants or bulbs bloom at a time that is not natural for them to do so.

genus Closely related plants grouped together under a single name, known as the "genus." Species are plants within a genus that can be separated from each other by recognizable individual characteristics. Each different plant is assigned a specific epithet. Taken together, the genus and specific epithet form the species name.

germination Growth of a plant embryo or the sprouting of a seed.

habitat The region in which a plant is found growing wild.

humus The stable organic constituent of soil that persists after the decomposition process of organic materials.

insecticide Also called "pesticide," a substance that kills insects by poisoning, suffocating, or paralyzing them. Types include stomach poisons, contact poisons, and fumigants. Organic gardeners avoid chemical pesticides, but may use natural ones such as pyrethrum, which is made from plant flowers.

legume A plant that is characterized botanically by fruit called a "legume" or "pod." This includes alfalfa, clover, peas, and beans, which are associated with nitrogen-fixing bacteria that can take nitrogen from the air and fix it within the plant. This provides nitrogen to the soil when the plant is plowed under or rototilled to make a garden.

loam A soil type made up of a mixture of sand, silt, and clay with various subdivisions that include sandy loam to clay loam.

organic matter Compounds containing carbon. In general it is material that once had been living. Organic matter in soil has a high nutrient exchange and moisture retention capacity and improves soil structure.

peat moss Partially decomposed vegetative material from sphagnum moss, which is useful as mulch or a soil additive to increase moisture-holding ability and porosity to soil.

perennial A woody to herbaceous plant that lives from year to year, and the plant's life cycle doesn't end with flowering or fruiting. Valuable for gardeners because they grow and bloom every year without the need for planting more seeds or seedlings.

petal One of the showy, usually colorful portions of a flower.

petiole The stalk of a leaf.

pH A term that represents the hydrogen ion concentration by which scientists measure soil acidity. The pH acidity scale measures from 1 (acid) to 14 (alkaline), with 7 as a neutral point.

phosphorus An essential macronutrient for plant growing. Its major function is promoting root formation in plants.

pistil The female reproductive structure of a plant found in the flower.

pollen Minute grains that carry male reproductive cells. These are borne on anthers of the bloom.

potash Potassium or potash is another essential macronutrient of plants. It is important for plant maturity, flower development, and hardiness.

rhizome A horizontal underground stem that gives rise to roots and shoots.

rototilling Garden soil preparation done with a rototiller that turns the soil to create a fine seedbed and area for planting.

seed A fertilized, ripened ovule (egg) that can grow into a new plant.

sepal One part of a whorl of green leafy structures that is located on the flower stem just below the petals.

soil A natural layer of mineral and organic materials that covers the surface of the earth at various depths which support plant life. Soil is created by the action of climate, water, time, and the interactions of living organisms and microorganisms on the parent material.

species A group of individuals forming a subdivision of a genus with similar characteristics, but differing from the genus too slightly to form another genus.

stamen The reproductive organ of the "male" pollen-bearing flower, the top part of which is the anther.

stigma The terminal part of the reproductive organ of the "female" flowering plant that receives pollen.

terminal bud The bud that is borne at the tip of a stem.

tuber A swollen, underground stem modified to store large quantities of food for the plant.

variety A group of individual plants, herbs, etc., forming a subdivision of a species with similar characteristics, but differing from the species too slightly to form another species.

Bibliography

Let's promote the planting and cultivation of more Biblical flower gardens across America and around the world. Not just herbs but also Biblical gardens with an entire range of plants of the Scriptures and Holy Land. The more you read and learn about this fascinating field of gardening, the easier it will be for you and your friends to become Biblical gardeners and apostles for Biblical gardens in your area.

My favorite reference books are at the top of this list, mainly because they are available today from bookstores, Internet sources, and some from out-of-print book specialists. Other reference books are mostly out of print but can sometimes be found through state library search systems, horticultural libraries, or rare-book dealers.

MOST AVAILABLE BIBLICAL PLANT RESOURCE BOOKS

Hareuveni, Nogah. *Ecology in the Bible,* 1974, 52 pages, 62 color photos.

———. *Nature in Our Biblical Heritage,* 1980, 146 pages illustrated.

———. *Tree and Shrub in Our Biblical Heritage,* 1984, 146 pages illustrated.

———. *Desert and Shepherd in Our Biblical Heritage,* 1991, 160 pages illustrated.

Hepper, F. Nigel. *Baker Encyclopedia of Bible Plants, Flowers, Trees, Fruits, Vegetables, Ecology,* 1992, 190 pages, illustrated.

———. *Planting a Bible Garden,* 1987 and 1997, 92 pages, illustrated.

Hudson-Knapp, M. *Plants in a Biblical Garden, Prayer Guide to the Children's Vegetable and Herb Garden.* A Biblical Garden Database via hkfamily@sover.net

King, E. A. *Bible Plants for American Gardens,* 1941 and 1975 revised edition.

Moldenke, H. N. *Plants of the Bible,* 1940, 135 pages.

Moldenke, H. N., and A. L. E. *Plants of the Bible,* 1952, 328 pages, reprint, 1986.

Swenson, Allan A. *Your Biblical Garden,* Doubleday edition, 1981, 220 pages.

———. *Plants of the Bible and How to Grow Them,* 1995, 220 pages.

Zohary, Michael. *Plants of the Bible,* 1982, 220 pages.

OTHER BIBLICAL PLANT RESOURCE BOOKS

Alon, Azaria. *The Natural History of the Land of the Bible,* 1969, 276 pages.

The American Horticultural Society's A–Z Encyclopedia of Garden Plants.

Crowfoot, G. M., and L. Baldensperger. *From Cedar to Hyssop: A Study in the Folklore of Plants in Palestine,* 1904, 204 pages.

Zohary, Michael. *Flora of the Bible, Interpretors Dictionary of the Bible,* 1962.

———. *The Plant Life in Palestine,* 1962.

BIBLES FOR BIBLICAL SCRIPTURE COMPARISONS AND PLANT IDENTIFICATIONS

Bible, Authorized King James Version, translated out of the original tongues and with the former translations diligently compared and revised, Judson Press, 1942.

Bible, English. Holy Bible, Authorized King James Version, 1967.

Bible, English. Revised Standard Version translated from the original languages as the version set forth in 1611, revised 1946–1952 and 1971.

Bible, Good News for Modern Man, The New Testament in Today's English Version, American Bible Society, 1966.

Bible, Life Application Study Bible, New International Version, Tyndale House Publishers, and Zondervan Publishing House, 1997.

Bible, The New English Bible, Standard Edition, Oxford University and Cambridge University Press, 1970.

Bible – Various Translations, Versions at Bible Study Tools website www.biblestudytools.com.

Cruden, Alexander, MA. Cruden's Concordance. Remains a valuable guide to the Scriptures of the Old and New Testaments of the King James Version.

Goodspeed, Edgar J. Popular Edition, An American Translation, 1935.

Moffatt, J. The New Testament: A new translation together with the Authorized Version. Parallel edition with introduction, 676 pages, 1922.

———. The Old Testament: A new translation. Volume I, Genesis to Esther, 571 pages, 1924. Volume II, Job to Malachi, 482 pages, 1925.

Phillips, J. B. The New Testament in Modern English, 1960.

Young, Robert, LL.D. Young's Analytical Concordance. This concordance organizes plants and other words based on their Greek and Hebrew names.

Index

Abies excelsa, 26
Abraham, 5
Absinthe, 41
Absinthin, 41, 42
Accho, 3
Acid soil, 123–24, 201
Adelman, Mrs. David C., 157–58
Adler, Betty, 158
Agave Americana, 11
Ahab, 6
Ail civitte, 82
Air freshener, rosemary and, 106
Akos, 51
Alexander the Great, aloes and, 10
Alkaline soil, 123–24, 201
Allium ampeloprasum, 56–58
Allium cepa, 66–69
Allium porrum, 56–57
Allium sativum, 53–56
Allium schoenoprasum, 81–83
Aloe africana, 10
Aloe arborescens, 10
Aloes, 8–11
Aloe succotrina, 9–10
Aloe vera, 8–11
Aloe vera chinensis, 10
Ambrosia melons, 65

Amenhotep III, 2
American Friends of Neot Kedumim, 170–71
Anemones, 152–53, 163
Anethum graveolens L., 20–23
Anise, 74–75
Annuals, defined, 201
Anthemis nobilis, 11–13
Anthers, defined, 201
Antique Flowers, 196
Aphrodisiacs
   coriander and, 16
   lavender and, 90
Aphrodite, marjoram and, 31
Apple mint, 98
Aquavit, 79–80
Archemorus, 102
Aristotle, 109
Artemesia absinthium, 39–42
Artemesia herba alba, 41
Artemesia judaica, 41
Auilaria agalocha, 9
Avatiach, 63
Averbach, Stan, 137–39, 156

Bacteria, defined, 201
Baked potatoes, chives and, 82

Balm, 75–77
Barlow, Marilyn, 191–92
Barrach, 78
Batavian escarole, 46–47
Bathing balms, rosemary and, 106
Beardsley, Mrs. George, 146
Beer, dandelion, 52
Bees, dandelions and, 51
Belgian endive, 46–47
Bermuda onions, 69
Beta carotene, lettuce and, 61
Bible Garden Links for Adults, 186
Bible translations. *See* Translations of Bible
Biblical Garden Fund, 138
Biblical Garden References, 186
Biblical garden team, 131–32, 135
Biblical Landscape Reserve (Israel). See Neot Kedumim
Biblical Plant Celebration Service, 174–83
   flowers, 178–79
   fruits, 181
   herbs, 179–80
   trees, 181–83
   vegetables, 179
Bibliography, 205–7
Biennials, defined, 201
Bitter herbs, 43–69
   chicory, 44–47
   cucumbers, 47–50
   dandelions, 50–53
   garlic, 53–56
   leeks, 56–58
   lettuce, 58–61
   melons, 61–66
   onions, 66–69
Black cumin, 18
Black mustard, 99, 100–101
Blood meal, defined, 202
Blood pressure, marigolds and, 103–4
Blooming times, 134
Bluestone Perennials, 189, 197
B'nai Shalom Biblical Garden (Walnut Creek, CA), 147–48
Boonstra, Sarah, 189
Borage, 77–79

Borago officinalis, 77–79
Boswellia carterii, 25
Boswellia papyrifera, 25
Boswellia sacra, 23–26
Boswellia thurifera, 25
Brassica nigra, 99–101
Bread, caraway, 80
Breath freshener, parsley and, 102
Brinkman Country Gardens (Canada), 186
Buckfast Abbey (England), 74
Buckland Abbey (England), 74
Buds, defined, 202
Bulb flowers, 178–79
   resources, 191, 192–93
Bulbs, defined, 202
Burgess Seed/Plant Company, 197
Burns, Betty, 156
Burpee, 196, 197
Burpee Hybrid melons, 65
Bush Crop cucumbers, 48
Buttercrunch lettuce, 59
Buttercups, 163
Butterhead lettuce, 59–61

Calcium, 124
   parsley and, 102
Calendula officinalis, 103–5
Canaanites, 2
Candy, horehound, 94, 95
Cantaloupes, 62–63
Capers, 27
Capparis sicula, 27
Caraway, 16, 17, 79–81
Carum carvi, 79–81
Cathedral Church of St. John the Divine Biblical Garden (New York City), 144–45, 186
Cedar of Lebanon, 163
Chamaemelum nobile, 11–14
Chamomile, 11–14
Charlemagne, 44, 85, 93
Chaucer, Geoffrey, 97
Chemical fertilizers, 202
Chewing gums, mint and, 97
Chicory, 44–47

Children, involving in Biblical garden, 133–34, 135–36
Chinese, ancient
  coriander and, 15
  garlic and, 54–55
  onions and, 67
Chives, 81–83
Church of St. John the Divine Biblical Garden (New York City), 144–45, 186
Church of the Holy Spirit Biblical Garden (Orleans, MA), 145
Church of the Wayfarer Biblical Garden (Carmel-by-the-Sea, CA), 146
Church service, 174–83
  flowers, 178–79
  fruits, 181
  herbs, 179–80
  trees, 181–83
  vegetables, 179
Church suppers, 167, 168
Cichorium endivia, 46
Cichorium intybus, 44–47
Cimmaron lettuce, 59
Citrullus lanatus, 47
Citrullus vulgaris, 61–66
Clement, Betty, 154
Clyde Robin Seed Company, 196, 197
Cochlearia armoracia, 87–90
Coffee
  chicory and, 45
  dandelion and, 52
Colds, mustard and, 100
Colic, lovage and, 93
Collingwood Ingram rosemary, 107
Comfrey, 84–85
Commiphora kataf, 35
Commiphora myrrha, 33–36
Community outreach, 133, 138–39
Compost (composting), 117–19
  defined, 202
  field, 119
  moisture and, 118–19
Concordia Lutheran Church Biblical Garden (Manchester, CT), 146–47
Condo communities, container gardens and, 130

Congestion, mustard and, 100
Congregation B'nai Shalom Biblical Garden (Walnut Creek, CA), 147–48
Container gardens, 126–30
  combination of plants, 127
  drainage of, 128
  fertilizer for, 129–30
  planting, 126–27
  soil for, 127–28
  watering, 128–29
Cook's Garden, 196, 197
Coriander, 14–17
Coriandrum sativum, 14–17
Corinthians, 131
Corms, defined, 202
Cornish lovage, 92–94
Cosmetics, lavender and, 90
Cough drops and syrup, horehound and, 95
County extension agents, 123, 124, 125
Crisphead lettuce, 59–61
Crocuses, 111–12
Crocus sativus, 5, 110–12
Crossman, Ken, 161
Crystal Palace Perennials, 197
Crystal White Wax onions, 69
Cucumbers, 47–50
Cucumis cantalupensis, 62–63
Cucumis melo, 47, 61–66
Cucumis sativa, 47–50
Culpeper, Nicholas, 73–74
  balm and, 76
  chicory and, 44
  fennel and, 85
  hyssop and, 28
  loveage and, 93
  mint and, 97
  rue and, 109
  savories and, 113
Cultivars, defined, 202
Cumin, 17–20
Cummum cyminum, 17–20
Curly leaf parsley, 102–3
Curly mint, 98
Curry powders, cumin and, 18

Index

Dale of the Song of Songs (Neot Kedumim, Israel), 169–70
Dalmatia, 37
Dandelions, 50–53, 148–49
David, 4
Dead Sea, 4
Deborah, 2
Decomposition, 202
De Materia Medica (Dioscorides), 73
Depression, rosemary and, 106
Desert and Shepherd in Our Biblical Heritage (Hareuveni), 165, 206
Deuteronomy, xii, 6, 39, 163
Dietary fiber, lettuce and, 61
Dijon mustard, 100
Dill, 16, 20–23
Dilla, 21
Dioscorides, 73
  borage and, 77
  chamomile and, 12
  chicory and, 44
  comfrey and, 84
  lemon balm and, 76
  oregano and, 101
Drake, Francis, 74
Dried flower arrangements, chamomile and, 12, 14
Drip irrigation, defined, 202
Drip Rite Irrigation, 196, 197
Dropsy, hyssop and, 28
Dry lawn clippings, 118, 120
Dugdale, Amy Tudor, 74
Dutch crocus, 112
Dutch Gardens, 189, 197

Eaglewood, 9
Early Bronze Age, 1
Ed Hume Seeds, Inc., 197
Edward E. Kahn Memorial Biblical Garden (Newport News, VA), 159
Egyptian onions, 66–69
Egyptians, ancient
  chicory and, 44
  coriander and, 15
  cumin and, 18
  garlic and, 54

  horehound and, 95
  muskmelon and, 62
  myrrh and, 34–35
  onions and, 67
Ein Karem (Israel), 171
Elizabeth, Queen of Hungary, 106
Elizabeth I, Queen of England, 45, 74, 80
Embalming
  aloes and, 9
  myrrh and, 34, 35
Endive, 45–47
English Physician and Complete Herbal (Culpeper), 74
Escarole, 46–47
Esob, 27–28
Esther, myrrh, 33
Ethiopian cumin, 18
Evangelism themes, 132
Exodus, xii
  bitter herbs, 43, 71
  coriander, 14
  frankincense, 23
  hyssop, 26
  marjoram, 31
  moriah, 165–66
  sage, 36, 165–66
Ezekiel, 2, 70

Fainting spells, lavender and, 90
Fair Haven Biblical Garden (Vermont), 131–32, 136–37, 148–49
  Web site, 140, 148, 186
Fall crocus, 112
Fennel, 85–87
Fertilizers, 121–23
  for container gardens, 129–30
  moderation in, 122–23
Field composting, 119
Finger, Martha, 158
First Biblical Resources U.S.A., 186
First Congregational Church Biblical Garden (Winter Park, FL), 161
First Presbyterian Church Biblical Garden (Nashville, TN), 150, 187
Flat leaf parsley, 102–3
Fleas, rue and, 109

Index

Flowers. *See also* Bulb flowers
   church service reflection about Biblical, 178–79
Flowery Branch Seeds, 198
Foeniculum vulgare, 85–87
Foliar fertilization, defined, 202
Food-preservation methods, xiii
Forcing, defined, 202
Fordhook lettuce, 59
Forest Farm Nursery, 189, 198
Frankincense, 23–26
Freedman, Janet, 158
French lavender, 91–92
Frenkley, Helen, 163
Frisan endive, 46
Froehl, Mary Lou, 161
From Garden to the Kitchen, 196, 198
Fruits, church service reflection about Biblical, 181
Funding (fund-raising), 137–39
   church suppers, 167, 168
Funugreek, 57

Gad, 15
Gail, Peter. A., 52–53
Galbanum, 86
Gardener's Supply Company, 195, 196, 198
Garden of Gethsemene, 140, 171–72
Gardens Alive, 196, 198
Garlic, 53–56
Genesis, 5, 70, 140
   creation, xi, 1, 174
Genus, defined, 202
Gerard, John, 73–74
   borage and, 77
   calendulas and, 104
   mints and, 97
   oregano and, 101
German chamomile, 13–14
Germination, defined, 202
Giant Red Hamburger onions, 69
Gibson, Mary Jo, 161
Glossary of terms, 201–4
Goldberg, Sol, 136
Goldfarb, Lee Kahn, 159
Good Deed Days, 138–39

Grapes, 181
Great Dandelion Cookbook (Gail), 52
Great Lakes lettuce, 59
Greenhart lettuce, 59
Green mint, 96–99
Green Towers lettuce, 59
Grieve, Maude, 19, 21, 90
Grosse Bouclee escarole, 46
Growing tips
   aloe, 11
   anise, 75
   balm, 76–77
   borage, 78–79
   calendulas, 105
   caraway, 81
   chamomile, 13–14
   chicory, 45–46
   chives, 83
   comfrey, 85
   coriander, 16–17
   crocus, 112
   cucumbers, 49–50
   cumin, 20
   dandelion, 53
   dill, 22–23
   fennel, 86–87
   garlic, 55–56
   horehound, 95–96
   horseradish, 89–90
   hyssop, 29–30
   lavender, 92
   leeks, 57–58
   lettuce, 60–61
   lovage, 94
   marjoram, 32–33
   melons, 64–66
   mints, 97–98
   mustard, 100–101
   myrrh, 36
   onions, 68–69
   parsley, 103
   rosemary, 107–8
   rue and, 110
   sage, 39
   savory, 114–15
   wormwood, 41–42

Habitat, defined, 202
Hadley, Arthur, 160–61
Hallowell, Steve, 139, 185
Hareuveni, Ephraim and Hannah, 162, 164–65, 166
Hareuveni, Nogah, 164–65, 205–6
Harris Seeds, 196, 198
Harvesting herbs, 72
Hastie, Drayton, 151
Hatfield House (England), 74
Headaches
  chamomile and, 12
  lovage and, 93
  rosemary and, 106
Heart disorders, balm and, 76
Hearts of Gold melons, 65
Henry Doubleday Association, 84
Henry Field's, 198
Henry IV (Shakespeare), 99
Hepper, F. Nigel, 86, 206
Herb, use of term, xii–xiii
Herball or General Historie of Plantes (Gerard), 74
Herba Taraxacon, 51
Herbs of the Bible (Swenson), 184–200
Hercules, 102
Hermes, 96
Herodotus, 24
Highlands Presbyterian Church Biblical Garden (Gainesville, FL), 150–51, 187
Hippocrates, 15
Hittites, 2
Holy Land
  climatic conditions, 3–4
  geography, 3–4
  history of, 1–3
Holy Spirit Biblical Garden (Orleans, MA), 145
Homer, *Odyssey*, 54, 76
Honey-mustard vinegrette, 100
Honey Rock melons, 65
Horehound, 94–96
Horehound candy, 94, 95
Horseradish, 87–90

Horus, 95
Howard, Albert, 117–18
Hudson-Knappp, Marsh, 148–49, 160, 161, 206
  on garden as focal point, 131–32
  on involving children, 135–36
  on Passover, 43–44
  on wormwood, 42
Humus, defined, 203
Hungarian goulash, caraway and, 79
Hunt, Mary, 152
Hussopos, 28
Hyssop, 4, 26–30
  growing tips, 29–30
Hyssopus officinalis, 26–30

Iceberg lettuce, 59–61
Incense, frankincense and, 24
Indigestion
  coriander and, 15
  lavender and, 90
  rue and, 109
Indoor herbal gardens. *See* Container gardens
Indore composting, 117–18
Insecticide, defined, 203
International Horseradish Festival (Collinsville, IL), 88
Iron, parsley and, 102
Iron Age, 2
Isaiah, 169
  chamomile, 11
  cucumbers, 47
  cumin, 17, 18
  dill, 20
Isothiocyanate, 88
Israel, history of, 1–4
Italian lovage, 92–94

Jacob, 5
Jacob, Irene, 154
Jaundice
  hyssop and, 28
  marigold and, 104
Jefferson, Thomas, 44, 76
Jeremiah, wormwood, 40

Jesus, 132, 140–41, 149
  aloes and, 9
  Garden of Gethsemene and, 140, 171–72
  mustard and, 99
  myrrh and, 5
J.L. Hudson Seedsman, 190, 198
Joel, 70
John, 4, 70
  aloes, 8, 9
Johnny's Selected Seeds, 189, 196, 198
Joseph of Arimathea, 140–41
Jossely, John, 113–14
Jotham, 4
Judean sage, 36–39
Judges, 4

Kammon, 18
Karkom, 110
Kasanof, Innes, 170–71
Kearns, Pud, 161
Ketzah, 18–19
Kings I, xii, 3, 6, 70
  hyssop, 26
  marjoram, 30
Kitchen herb garden, 6, 70–115
  anise, 74–75
  balm, 75–77
  borage, 77–79
  caraway, 79–81
  chives, 81–83
  comfrey, 84–85
  fennel, 85–87
  horseradish, 87–90
  lavender, 90–92
  lovage, 92–94
  marrubium, 94–96
  mints, 96–99
  mustard, 99–101
  oregano, 101
  parsley, 101–3
  pot marigold, 103–5
  rosemary, 105–8
  rue, 108–10
  saffron, 110–12

savory, 112–15
Klehm's Song Sparrow Perennial Farm, 198
Kligfeld, Bernard, 158

Lactuca sativa, 58–61
Lamb mint, 96–99
Lamentations, wormwood, 40
Languis, Marvin, 160
Lavandula augustifolia, 90–92
Lavandula stoechas, 91
Lavandula vera, 91
Lavender, 90–92
Lavender Lady, 92
Leaf lettuce, 58–61
Leeks, 56–58
Legumes, defined, 203
Lemon balm, 76–77
Lettuce, 58–61
Levisticum officinale, 92–94
Leviticus, 4
  frankincense, 23
  hyssop, 26–27
  marjoram, 30
Lily of the field, 149, 152–53
Lilypons Water Gardens, 196, 198
Lime, 124–25
Linnaeus, 75–77, 88
Liver damage, comfrey and, 84
Loam, defined, 203
Location of herb garden, 116
Loosehead lettuce, 59–61
Lovage, 92–94
Luke
  frankincense, 23
  rue, 108

McGarity, Jan, 147
McMahon, Page, 145
Magnolia Plantation Biblical Garden (Charleston, SC), 151, 187
Mail-order catalogs, 139, 188–200
Mailorder Gardening Association (MGA), 194–95, 196, 200
Majorana syrtacu, 27

Majorca Pink rosemary, 107
Mandell Hy, 158
Mantis, 198
Manure, compost and, 118
Marigold, 103–5
Marjoram, 28, 30–33
Mark, 5, 171–72
Marrob, 94–95
Marrubium, 94–96
Marrubium vulgare, 94–96
Matricaria recutita, 13–14
Matthew, xiii
  anise, 74
  cumin, 17–18
  frankincense, 23, 25
  mustard, 99
  myrrh, 34
Meadowsweet Herb Farm, 107, 190
Medicinal value of herbs, 71–72
Melissa officinalis, 75–77
Mellinger's, 190, 198
Melons, 61–66
Menorah, moriah and, 165–66
Mentha piperita, 96–99
Mentha pulegium, 98
Mentha spicata, 98
Mentha suaveolens, 98
Mentha viridis, 96–99
Merenptah, 2
Mesclun mix greens, 61
Middle Kingdom, 2
Miller Nurseries, 190, 198
Milton, John, *Paradise Lost*, 85–86
Mints, 96–99
Mint sauce, 98
Miracle-Gro, 122
Modern Herbal, A (Grieve), 19
Moffatt, James, 111
Moldenke, Alma L., 15–16, 206
Moldenke, Harold N., 15–16, 18, 206
Mor, 5
More, Thomas, 106
Morgan, Bob, 152–53
Moriah, 165–66
Moses, 2, 48, 155, 163, 165–66
  myrrh and, 34–35

Mountain radish, 87–90
Mountain States Wholesale Nursery, 159
Mountain Valley Growers, Inc., 190, 196
Mount Gerizim, 4
Mount Herman, 3–4
Mount of Olives, 171–72
Moutarde des Allemands, 87
Mulch (mulching), 119–20
Muscle problems, rosemary and, 106
Muskmelon, 61–66
Mustard, 99–101
Mustard plasters, 100
Myrrh, 4–5, 25, 33–36
  growing tips, 36
Myrrha, 34

Naboth, 6
Narcissus, 163
Nardus Italica, 91
National Gardening Association (NGA), 195, 196
Nature in Our Biblical Heritage (Hareuveni), 165, 205
Nehemiah, 2
Nelson, Taylor Drayton, 151
Neot Kedumim (Israel), 162–71
  educational programs, 166
  general information, 170–71
  herb guide, 164
  meals and menus, 166–68
  Web site, 187, 195–96
  Wedding Trail, 168–69
Nero, 56
New Kingdom, 2
Nichols Garden Nursery, 196, 199
Nicodemus, 9
Nigella sativum, 18
Nitrogen, 118, 121–22
Northwoods Nursery, 199
Numbers, xii, 70, 71
  bitter herbs, 43
  coriander, 14
  hyssop, 26, 27
  melons, 61
  onions, 66
Odyssey (Homer), 54, 76

Ojai Presbyterian Church Biblical Garden (CA), 152
Old English lovage, 92–94
Old House Gardens, 191, 196
Olive trees, 165, 171–72, 181
One Green World, 199
Onions, 66–69
Oregano, 101
Organic Gardening, 196, 197
Organic matter, defined, 203
Origanum maru L., 30–33
Origanum onites, 32
Origanum syriacum, 27, 30–33
Origanum vulgare, 32, 101
Our Lady's mint, 96–99
Ovid, 44

Palestine, history of, 1–4
Paradise Lost (Milton), 85–86
Paradise Valley United Methodist Church Biblical Garden (AZ), 152–53, 187
Paris Island lettuce, 59
Parkinson, John, 89, 113
Park Seed, 191, 197, 199
Parkside Lutheran Church Biblical Garden (Buffalo, NY), 161
Parsley, 101–3
Passover, 4, 88
  bitter herbs and, 43–44, 148–49
Paver kits, 138
Peat moss, defined, 203
Pennyroyal, 98
Peppermint, 96–99
Perennials, defined, 203
Perfume, lavender and, 90, 91
Petals, defined, 203
Petiole, defined, 203
Petit poureau, 82
Petroselinum crispum, 101–3
Philistines, 2
Phosphorus, 121–23, 124, 203
pH scale, 123–24, 203
Pickling
  cucumbers, 47, 48
  dill and, 21, 22

Pimpinella anisum, 74–75
Pine bark, 120
Pistil, defined, 204
Plants of the Bible (Moldenke), 15–16, 206
Pliny the Elder, 72–73
  anise and, 74
  balm and, 75
  borage and, 77
  chamomile and, 12
  chicory and, 44
  comfrey and, 84
  cucumbers and, 48
  fennel and, 85
  garlic and, 54
  horseradish and, 88
  mints and, 96
  oregano and, 101
  rue and, 109
  savories and, 113
Pollen, defined, 204
Pomegranates, 168
Potash, defined, 204
Potassium, 121–23
  lettuce and, 61
Pot marigold, 103–5
Pot marjoram, 32
Potpourri
  chamomile and, 12, 14
  lavender and, 90
  rosemary and, 106
Prince of Peace Lutheran Church Garden (Augusta, ME), 153
Prizehead lettuce, 59
Prostratus rosemary, 107
Proverbs, 167
  aloes, 9
  wormwood, 40
Psalms, xiii, 4, 163, 168, 170
  aloes, 9
  hyssop, 27, 28
  myrrh, 33
Pyramid Age, 1–2
Pythagoras, 74

Quality Dutch Bulbs, 191, 199

Rameses II, 2
Raphanos agrios, 88
Red cole (horseradish), 87–90
Red Fire lettuce, 59
Red Sails lettuce, 59
Redwood chips, 120
Reference books, 205–7
Resources, 184–200. *See also* Web sites
Reuo, 108
Rhizomes, defined, 204
Richters, 191
Robinson, E. Lamar, 175
Rodef Shalom Biblical Botanical Garden (Pittsburgh, PA), 153–54, 187
Romaine lettuce, 59–61
Roman chamomile, 11–14
Romans, ancient (Roman Empire), 2–3
   borage and, 77
   chives and, 82
   cucumbers and, 48
   cumin and, 19
   fennel and, 85
   frankincense and, 24
   garlic and, 54
   horehound and, 94
   lettuce and, 58
   marigold and, 104
   marjoram and, 31
   mints and, 96
   muskmelon and, 62
   mustard and, 99
   parsley and, 102
   rue and, 108
   savories and, 113
Roris Gardens, 199
Rosemary, 105–8
Rosmarinus officinalis, 105–8
Rototilling, defined, 204
Royal River Roses, 199
Ruby lettuce, 59
Rue, 108–10
Ruta graveolens, 108–10

Sadler, Frances, 158
Saffron, 5, 110–12
Sage, 36–39, 165–66
Sagebrush, 41. *See also* Wormwood
Sage of Bethlehem, 96–99
St. Gregory's Episcopal Church Biblical Garden (Long Beach, CA), 154
St. James Lutheran Church Biblical Garden (Coral Gables, FL), 155
St. James Lutheran Church Biblical Garden (Wapwallopen, PA), 161
St. John's Episcopal Church Biblical Garden (Norman, OK), 155–56, 187
St. John's Episcopal Church Biblical Garden (Worthington, OH), 160–61
St. John the Divine Biblical Garden (New York City), 144–45, 186
Salanca escarole, 46
Salvere, 37
Salvia judaica, 36–39
Salvia lyrala, 38
Salvia officinalis, 36–39
Salvia salvatrix, 37
Salvia urticifolia, 38
Samson melons, 65
Sandalwood, 9
Santalum album, 9
Satan, garlic and, 54
Satureja hortensis, 112–15
Satureja montana, 112–15
Saul, 2
Savories, 112–15
Schultz fertilizers, 122
Scott, Joseph, 131, 153
Sea of Galilee, 3, 4, 182
Seder, 43–44, 148–49
Seeds, defined, 204
Seeds of Distinction, 192, 199
Select Seeds, Antique Flowers, 191, 199
Sepal, defined, 204
Shakespeare, William
   caraway and, 80
   marigold and, 104
   mustard and, 99
   rue and, 109
   savory and, 113
Shallots, 82
Sherry wine, chamomile and, 12

Shir Ami Biblical Garden (Newtown, PA), 156
Sidell, Shirley Pinchev, 139–40, 147–48, 185
Signs, for Biblical gardens, 135
Skin disease, hyssop and, 4
Skin problems, comfrey and, 84
Soaps, mint and, 97
Soil, 116–17, 123
  composting, 117–19
  for container gardens, 127–28
  defined, 204
  fertilizers, 121–23
  lime, 124–25
  pH of, 123–24, 203
Solomon, 4, 111, 169
Song of Solomon, xiii, 4, 5, 70, 155, 163, 169–70
  myrrh, 33
  saffron, 110, 111
Sore throats
  horehound and, 95
  hyssop and, 28
Sorghum, 27
  Sorghum vulgare, 27
Sourby, Charles, 140–42
Southern exposures, 116
Spartan Valor cucumbers, 50
Spearmint, 96–99
Special Celebration of Plants of the Bible, 176–83
  flowers, 178–79
  fruits, 181
  herbs, 179–80
  trees, 181–83
  vegetables, 179
Species, defined, 204
Spike lavender, 91
Spike oil, 91
Stamens, defined, 204
Stanislaus I, King of Poland, 66
Stigma, defined, 204
Stokes Seeds, 199
Stokes Tropicals, 197, 199
Strybing Arboretum Biblical Garden (San Francisco, CA), 157

Sumerian Age (Sumerians), 2, 67
Summer Bibb lettuce, 59
Summer savory, 112–15
Sun City Nursery, 159
Sun exposure, 116
Sweet marjoram, 32
Sweet'n Early melons, 65
Sweet Spanish onions, 69
Symphytum officinale, 84–85

Talmud, 139
  coriander and, 15
  garlic and, 53–54
  saffron and, 111
Taraxacum officinale, 50–53
Taraxos, 51
Taylor, C. Powers, 144
Tea
  borage, 78
  chamomile, 12
  dandelion, 51
  horehound, 95
  hyssop, 28
  sage, 37–38
Teitlebaum, Shlomo, 169, 170
Temple Beth-El Biblical Garden (Providence, RI), 157–58
Temple Beth Shalom Biblical Garden (Sun City, AZ), 158–59, 187
Temple Sinai Biblical Garden (Newport News, VA), 159, 187
Tennyson, Alfred Lord, 178
Terebinth, 25
Terminal bud, defined, 204
Territorial Seed Company, 192
Themes, for Biblical gardens, 132, 134
Thesis, 34
Thompson, Mrs. Henry "Betty," 144
Thompson & Morgan, 197
Thutmose, 2
Tiberius, 48
Toothaches, mustard and, 99
Toothpastes, mint and, 97
Torah, 158–59
Tosca endive, 46
Translations of Bible, 19, 188, 207

Translations of Bible (*cont.*)
  aloe, 9
  cucumbers, 47
  cumin, 18–19
  dill, 20, 21
  hyssop, 27–28
  saffron, 111
Tree and Shrub in Our Biblical Heritage (Hareuveni), 165, 205
Trees, 141
  church service reflection about Biblical, 181–83
Trigonella foenum-graecum, 57
Tubers, defined, 204
Turner, William, 109

USDA (U.S. Department of Agriculture), 125, 134

Van Bourgondien Bulbs, 192, 197, 199
Variety, defined, 204
Vegetables, church service reflection about Biblical, 179
Vessey's Seeds, Ltd., 199
Victor, Mrs. Alexander O., 144
Vinegar, savory and, 113
Virgil, 44, 54, 113
Virgin Mary, 105, 141
Visser, Joop, 189
Vitamin A
  lettuce and, 61
  parsley and, 102
Vitamin B12, comfrey and, 84
Vitamin C, parsley and, 102
Volunteers, 135
Volunteer Vegetable Sampler (Gail), 52

Walnut trees, 169–70
Warsaw Biblical Garden (IN), 160, 188
Water gardens, 132, 136–37
Watermelons, 47, 61–66
  growing tips, 64–66
Wayside Gardens, 192, 199
Web sites, 139–40, 184–86
  Biblical gardens, 186–88
  plant sources, 188–200
Wedding Trail (Neot Kedumim, Israel), 168–69
Wegman, Jay, 145
Westlake, Grace, 150
White Flower Farm, 193, 197, 200
White horehound, 94–96
White lilies, 163
White marjoram, 27
Wild radish, 88
Wildseed Farms, Ltd., 197, 200
Wine, 181
  borage and, 78
  dandelion, 52
  lemon balm and, 76
Winter savory, 112–15
Winter's Tale (Shakespeare), 113
Wise Men
  frankincense and, 23, 24–25
  myrrh and, 34
Witloof chicory, 45–47
Wood, Tom, 160
Wood chips, 120
World of the Bible Gardens (Jerusalem), 171
Wormwood, 39–42

Yad-Hasmona Biblical Gardens (Israel), 188
Yellow Bermuda onions, 69
Yellow dye, saffron and, 110–11
Your Biblical Garden (Swenson), 151, 174, 206

Zeus, 96
Zohary, Michael, 18–19, 44, 206